HOUSE
BUILDING
MANUAL

Published by: National House Building Guarantee Company Ltd.,
Construction House,
Canal Road,
Dublin 6.

Authors: John A. McCarthy, B. Arch. MRIAI,
Coll + McCarthy Architects

Eugene Farrell, BE, M. Eng. Sc., PhD., C. Eng., MIEI

Anthony McFeely, Coll + McCarthy Architects

Designed by: Robert Graham

Illustrations by: Anthony McFeely

Printed by: Kellyprint

ISBN 0 9523614 1 8

First published 1993
Re-printed 1996
Available from: HomeBond
Construction House,
Canal Road,
Dublin 6.

Note: All standards referred to in this publication were current at the time of printing.

Contents

Foreword

Foreword by *Mr. Joe Tiernan*
Chairman, HomeBond

As Chairman of the HomeBond I take great pride in introducing this excellent and original manual, because I believe it is a splendid publication. In the same way I take great pride in being a home builder. Like most young people, when I was growing up, I wanted to be involved in doing something positive and creative. It was that ambition which led me to become a home builder and Property Developer. Since then, I have always taken pride in building good homes for so many Irish people over the last quarter of a century. I know that these ideals of achievement are shared by, and are a powerful motivation for, my colleagues in the Home Building Industry. Truly, we have cause to be proud, our industry has produced all the fine homes you see throughout the country.

This is not always an easy task, ours is a difficult business involving great risks and substantial venture capital. Prudence and sound judgment are prerequisites for success as we deal with the multiplicity of activities which our role requires.

First and foremost there is the new home itself. Homes are much more complex buildings than is commonly realised. The construction involves the use of many materials, skills and services, which must be assembled into an integrated structure which is soundly built, well lighted, heated and ventilated and with proper access; this ensures a high quality final product.

There are other essential activities connected with the production of a home such as site selection, planning and design, the provision of road works and other services including landscaping, dealing with the local authorities and the professionals such as architects, solicitors, engineers and quantity surveyors. Finally, it is important not to neglect the marketing aspects including advertising and sales.

The decision to produce this site manual derives from our recognition of this complexity.

Every part of house construction is considered and each trade is followed through all stages of construction. I believe that the manual will become an indispensable reference on every building site and as a text book in our universities and colleges of higher education because it should form part of the foundation for the training of architects, engineers and craft people of the future.

Pride in our achievements has led home builders to keep looking for ways to improve our product. Today houses are far superior to those of even a few years ago and represent a better investment than ever.

HomeBond has played a significant role in this process. We must never stand still however, and it is my belief that this manual will add further to our capacity to serve our public even better. I am therefore delighted to introduce it and to commend it to all who are in any way involved in the business of producing good quality homes in Ireland.

The Board wishes to express its thanks to all those who have been connected in the preparation and publication of this manual.

Joe Tiernan
Chairman
HomeBond.

Preface

To write a Manual on house building proved to be a much more difficult task than anticipated when the venture started. As the Chairman states in his foreword homes are much more complicated structures than most people imagine. A great number of houses are built and certainly many aspects of construction are repeated time after time; but the variety of designs and materials used makes for complexities.

Then the rules keep changing. During the time this publication was being put together the Building Regulations came into force on 1st. June 1992 and that meant new requirements which had to be incorporated in the structure of the dwellings and the text of the manual. There are modifications made to materials, new types of construction are adopted, trends change in design - all impact on house building and need to be considered. As far as it is possible to do so these are covered in this Manual.

HomeBond and the Department of Environment together inspect houses all over the country. As this Manual goes to print HomeBond has 150,000 registered dwellings. These houses are all inspected and in the main they are found to be well built.

There are items of construction noted at inspection stage which are not fully understood by those involved in building, and it is the aim of this Manual to set down in clear terms good advice for house builders.

Houses are put together by various trades working on site with many different materials. They are built in ground which may vary in its geology over a few metres, both horizontally and vertically. There is always something to be learned and there are, unfortunately, times when something goes wrong. The foundation manual published by HomeBond is a book which has helped builders understand the problems they may encounter below ground level and how to deal with them. This Manual deals with the construction of a house from foundations to chimney pot.

In the acknowledgements a list is given of the people who have helped with advice and they are thanked.

At HomeBond Dr. Eugene Farrell is the staff member who took the responsibility of editing the book in collaboration with John McCarthy, the architect commissioned by HomeBond to produce the book. Also very much involved was John McCarthy's assistant Anthony McFeely who compiled the book and produced the illustrations. It is appropriate to congratulate all three for their dedication, hard work and expertise.

This publication is produced as general guidance to members and it is appropriate to indicate that anybody who takes on the responsibility of producing a dwelling must themselves ensure that it is built to correct standards and complies in every way with the Building Regulations and all other requirements that are in force and impact on it. Neither HomeBond, nor the Authors, nor anyone who has assisted in the production of this manual can accept any liability arising out of reliance on any aspect of this publication. Persons should always obtain professional advice for their specific situation and requirements.

HomeBond which was established in 1978 has become a most important force in the private sector of housing. The Board involved in producing this Manual is as follows:

Joe Tiernan (Chairman)
Michael Greene, B.L., B. Comm., M. Econ. Sc., (Managing Director and Secretary)
W. Brian Boyd, MRIAI, RIBA (General Manager)
Martin Browne, FCA
Michael J. Coleman
Frank Fahy
Frank McGee
Sean McKeon
Don O'Brien, FCA
Paddy Raggett
Francis Rhatigan
Jim Wood

It is the opinion of the Chairman and Board that the Manual will be invaluable for all concerned with good house building.

Acknowledgement

HomeBond gratefully acknowledges the assistance of the following people in the compilation of this publication. All sections of the industry and the persons contacted gave their advice and comments and without their help it would not have been possible to produce this publication.

John McCarthy, *Anthony McFeely.*	Coll + McCarthy Architects.
Brian Boyd, *Eugene Farrell,* *Tom Crotty,* *Kevin Dillon,* *Mike O'Grady,* *Tom Cregg,* *Kevin Mongey, R.I.P.*	HomeBond
Frank Lee, *Gerry McCabe.*	Lee McCullough & Partners Consulting Engineers.
Eoin O'Cofaigh.	McHugh O'Cofaigh Architects.
Sean Wiley, *Bill Robinson,* *Bob Davis,* *Glynn Douglas,* *Seamus Kelly,* *Bill Quinn.*	EOLAS
Brendan Finlay, *Kevin Spencer.*	Department of Environment.
James Reynolds, *Raymond Spain.*	Richmond (Irl) Bldg. Products.
Hugh Boyd, *Michael Weldon.*	Wavin (Irl) Ltd.
Domhnall Blair, *Tom Davis,* *Cyril Pearson.*	Roadstone Dublin Ltd.
Paul VanCauwelaert.	Roadstone Provinces Ltd.
Declan Grehan.	Tegral Bldg. Products.
George Yeates.	Gypsum Industries plc.
Sean Hyde.	Health & Safety Authority.
Colm Bannon.	Irish Cement Ltd.
Niall Walsh.	Dublin Corporation.
Nick Ryan.	Environmental Research Unit.
Joe O'Connor.	Bat Metalwork Ltd.
Michael Spillane.	Carey Glass Ltd.

WORK TO D.P.C. LEVEL INCLUDING GROUND FLOOR CONSTRUCTION

WHAT TO DO BEFORE BUYING A SITE

You should visit the site and adjoining areas to see exactly what you are buying. This will take no more than 2 or 3 hours of your time.

Then follow the 4 rules for site visits.

RULE 1
LOOK AT ADJOINING BUILDINGS.
Look for:

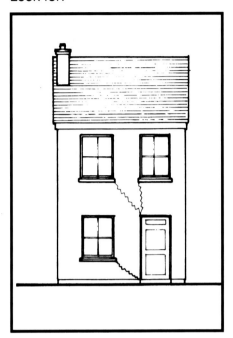

(A) Signs of damage—if so why?

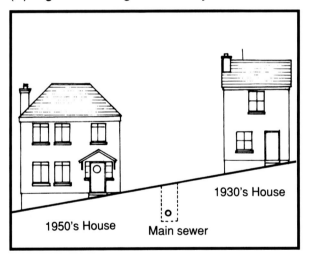

(B) Slopes—what effect? Gaps—why?

What effect will the house you are to build have on existing houses?
What effect will an existing house have on the one you are to build?
Why was the gap left between the houses?

RULE 2
LOOK AT PLACE NAMES.
What do they tell you?

WHAT TO DO BEFORE BUYING A SITE
continued.

RULE 3
Talk to local people.
"A few years back there was a dump in the area".

"..........and I know you WON'T believe me but we used to catch fish there before the council filled it in".

These simple precautions may give vital clues about what used to happen and what problems may be expected.

RULE 4
Walk through the site and immediately adjoining land.
Obtain the owner's permission to walk through the land.
To stand at the edge and look in or to drive by is not enough; walk right through and look carefully at:

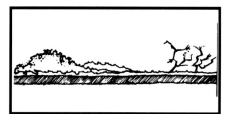

(A) Type of ground: Surface condition.

(B) Any open trench or working which shows ground strata.

(C) Vegetation; are there plants which would indicate marshy ground?

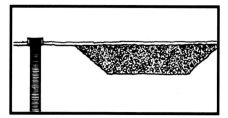

(D) Evidence of previous use of site: was it a filled site?

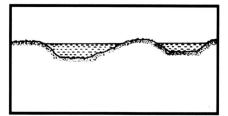

(E) Local rivers or streams: Possibilities of flooding. Also check ground water levels after heavy rain.

WHAT TO DO BEFORE BUYING A SITE
continued.

Even if you have a life-long knowledge of a site and are practically certain that it has no problems, it is still prudent to obtain the owner's permission to carry out a preliminary, low cost ground investigation before you buy.

An excavator will dig trial pits down to 4m in depth. This is normally sufficient to reveal most of the hazards relevant to 1,2 and 3 storey work.

Where a builder is tendering to build on a client's own site, it is imperative that ground investigation as outlined above is carried out. This will help avoid disputes over additional costs arising from unforeseen extras in foundation work.

NUMBER OF TRIAL PITS:
NO FIRM RULE CAN BE GIVEN ON THE NUMBER OF PITS YOU DIG. YOU MUST SATISFY YOURSELF THAT YOU HAVE RELIABLE INFORMATION AS TO WHAT IS HAPPENING UNDERGROUND ALL OVER THE SITE.

Take a site of 10 houses:
If the conditions vary greatly, ten pits may be required.
If they do not vary, four or even two may be enough.
One pit is almost never reliable, even for a one-off dwelling.

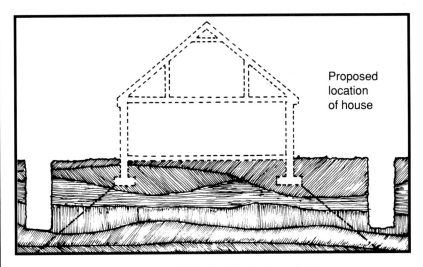

Proposed location of house

Location: Locate pits outside the likely foundation area to avoid creating soft spots.
Remember if foundations bear on different layers of subsoil, specialist foundation design may be required.

Shape: If the ground is found to be highly variable 50mm proving borings can be taken between trial pits and within the dwelling area.

SAFETY: APPROPRIATE SAFETY PRECAUTIONS MUST BE TAKEN PRIOR TO INSPECTION OF PITS. (See page 15)

WHAT TO DO BEFORE BUYING A SITE
continued.

Why has the site not been built on before?

Maps and photographs are a useful source of information about the characteristics and history of a site.

Even if you have a detailed, life-long knowledge of the site, it is wise to see what maps and photographs are available.

Ordnance maps.
Study the current map.
The question of why the site has not already been built on may be critical in an otherwise built up area. Look also at older Ordnance Survey maps as they may show ponds, hedges, buildings, ditches, rivers etc., now removed but which may have an effect on building.

O.S. maps are available from the Ordnance Survey Office, Phoenix Park, Dublin 8.

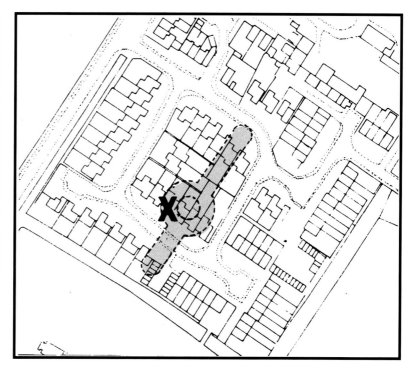

Example:
Settlement occurred in house at **X**. An old map showed that it had been built on the site of a water garden (shown dotted) from a previous era.

CHEMICAL HAZARDS AND CONTAMINANTS.

A building site may contain hazards and contaminants such as radon, methane, carbon dioxide, sulphates, chlorides or acids. These may be naturally occurring or may occur as a result of the previous use of the site.

This page and the following pages give outline guidance for dealing with such substances.

Radon:
Radon is a naturally occurring radioactive gas which may build up to unacceptably high concentrations in buildings. Areas of greatest incidence include all the counties of Munster and counties Galway, Mayo, Roscommon, Sligo, Louth and Wicklow.

Further guidance is given in Appendix A. Additional information is also available from:

- The Radiological Protection Institute of Ireland.
  3 Clonskeagh Square,
  Dublin 14.

- The Environmental Information Service (ENFO),
  17 St. Andrew's Street,
  Dublin 2.

- The Environmental Research Unit (ERU),
  St. Martin's House,
  Waterloo Road,
  Dublin 4.

- British Research Establishment (BRE) Report RB211:1991 "Radon Guidance on Protective Measures for new Dwellings".

- "Building on Derelict Land" by B. A. Leach and H. K. Goodger. CIRIA/PSA publication, ref: SP78, 1991.

Methane and Carbon dioxide:

Incidents involving landfill gas in buildings have increased in recent years. The principal components of landfill gas are methane (which is flammable) and carbon dioxide (which is toxic) and so if it enters the building it can pose a risk to both health and safety. These two gases are also associated with coal strata, river silt, sewage and peat.

Specialist advice should be sought, and site investigation carried out, if there is a risk of methane or carbon dioxide, and further information may be obtained from the British Research Establishment (BRE) report BR 212: 1991 "Construction of new buildings on gas-contaminated land" available from the British Research Establishment (BRE), Garston, Watford WD2 7JR, England.

Sulphates:
Sulphates can cause expansion and disruption of concrete, particularly on filled sites. Concrete mix specification should take account of the risk, and the concrete mix should be specified by a suitably qualified consulting engineer. Guidance is available in BRE Digest 363 "Sulphate and acid resistance of concrete in the ground".

Chlorides:
Chlorides increase the risk of reinforcement corrosion and chemical attack on concrete. As with sulphates, the concrete mix specification and reinforcement details where a chloride risk exists should be prepared by a suitably qualified consulting engineer. The engineer appointed should be qualified by examination, be in private practice and possess professional indemnity insurance.

CHEMICAL HAZARDS AND CONTAMINANTS
continued.

Acids:
High acid content, for example in peat, can damage concrete.

To avoid such damage precautions such as the following may be necessary:

(a) Increased cement in mix.
(b) Use of special cements.
(c) Thorough compaction.
(d) Use of a protective layer, such as bituminous or plastic membrane to prevent the contaminants coming into contact with the concrete.

Again, the advice of a suitably qualified person is essential to avoid costly repairs.

Additional information can be found in I.S. 326: 1988 "Code of Practice for the Structural Use of Concrete".

Contamination from a previous use of site:
Contamination arising from a previous use is a possibility requiring investigation on some sites. The table opposite gives some guidance to identifying such risks and possible contaminants.

Further guidance is set out in BS 5930: 1981 "Code of practice for site investigation" and DD 175: 1988 "Code of practice for the identification of potentially contaminated land and its investigation". Both these documents are published by the British Standards Institution (BSI), Linford Wood, Milton Keynes, MK 14 6LE, England.

Sites likely to contain contaminants

◆ Asbestos works.

◆ Gas works.

◆ Industries making or using wood preservatives.

◆ Landfill and other waste disposal sites or ground within 250 metres of such sites.

◆ Metal mines, smelters, foundries, steel works and metal finishing works.

◆ Oil storage and distribution sites.

◆ Paper and printing works.

◆ Railway land, especially the larger sidings and depots.

◆ Scrap yards.

◆ Sewage works, sewage farms and sludge disposal sites.

Identifying Contaminants

Signs of possible contamination	Possible contaminant
(a) Vegetation (absence, poor or unnatural growth)	Metals, metal compounds, organic compounds, gases.
(b) Surface materials (unusual colours and contours may indicate wastes and residues)	Metals, metal compounds, oily and tarry wastes, asbestos (loose), other fibres, organic compounds including phenols, potentially combustible material including coal and coke dust, refuse and waste.
(c) Fumes and odours (may indicate organic chemicals at very low concentrations)	Flammable, explosive and asphyxiating gases including methane and carbon dioxide, corrosive liquids, faecal, animal and vegetable matter (biologically active).
(d) Drums and containers (whether full or empty)	Various.

BUILDING ON HAZARDOUS GROUND.

Engage an engineer and have a thorough site investigation carried out by a site investigation specialist. The engineer can then design appropriate foundations based on the site investigation findings.

The engineer appointed should be qualified by examination, be in private practice, and possess professional indemnity insurance. The engineer's report may mean that it will pay the builder to use the site differently from his original idea.

Site layout
If any part of the site is hazardous, the best course may be to arrange the dwellings so that the hazardous area is not built on, for example:

(A) Resite dwelling.

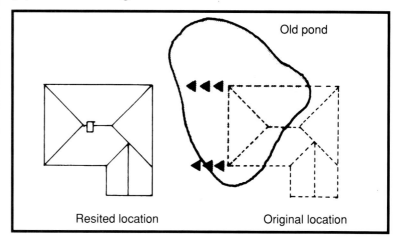

Resited location   Original location

(B) Leave an open space.

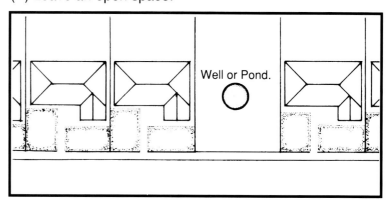

MAKE FINAL DECISION.

Length of Blocks.
If the problem is the likelihood of variable settlement, it may be prudent to change from long terraces with large differential movement, to semi-detached or detached houses with comparatively little differential movement.

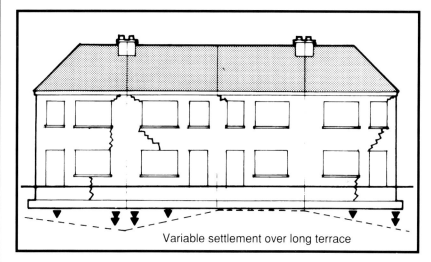

Variable settlement over long terrace

Problem: Variable settlement can cause cracking.

Ground conditions which are clearly poor require site investigation.

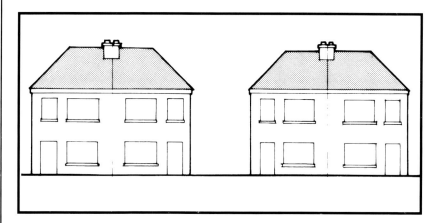

Design to suit conditions.
Risk of cracking reduced by the use of detached or semi-detached house.

INTRODUCTION

Reproduced on this page is the text of an HomeBond guidance note to members on the subject of vibratory ground improvement techniques.

Note:
This method of foundation construction may not suit all ground types and may generate more risk than other forms of ground improvement methods listed below. Builders should exercise caution and engage experts if using vibro compaction.

Other ground improvement techniques for hazardous ground include: bored piles, driven piles, ground beams and the use of suspended ground floors. All these techniques should be designed and supervised by an independent engineer, as defined elsewhere on this page.

Vibratory ground improvement techniques.

This is a system where stone columns are installed to improve the load bearing capacity, reduce settlement and provide an adequate bearing strata for foundations. In cases where the method is to be used by members, HomeBond outlines the following requirements that must apply.

An independent engineer must be engaged by the builder. The engineer appointed must be qualified by examination in civil or structural engineering, be in private practice and possess professional indemnity insurance.
The engineer will determine:-

A   The nature and extent of ground hazards.

B   The suitability, or otherwise, of vibratory ground improvement technique. This may be in conjunction with specialist contractor's engineer.

C   The suitability of dwelling layout, dwelling design and foundation design.

The builder should obtain written confirmation from the engineer and specialist contractor that the site is suitable for the system selected.

The engineer and contractor should agree who is responsible for procedures and testing and agree on the types and frequency of testing. The engineer should provide competent supervision. When tests have been carried out the engineer should confirm to the builder that they are satisfactory.

Aspects such as drainage layout, design of foundations, and design of suspended floors are the responsibility of the independent engineer.

The independent engineer should engage a specialist site investigation firm to carry out site investigations to assess site conditions and also to investigate for any gas content or harmful wastes.

Once the stone columns have been installed it is vital that drainage or other service trenches are not dug adjacent to the columns for such activity could remove lateral support to the columns.

HomeBond will require confirmation on completion both from the vibratory ground improvement contractor and the engineer that the foundations are satisfactory to ensure the structural stability of the dwelling taking into consideration the ground conditions that prevail.

WHERE TREES OCCUR ON SITE.

A

Determine existing root growth and check excavations carefully for roots and fibrous material. The presence of roots can cause movement of foundations and consequent structural damage.

Root fibres visible on sides and bottom of trench.

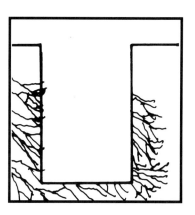

B

Assess likely pattern of future root growth.

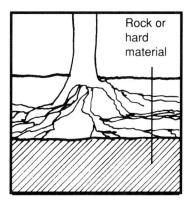

Roots are likely to remain above hard material such as rock.

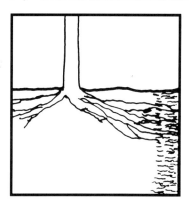

Roots are likely to grow a long way to reach a wet area.

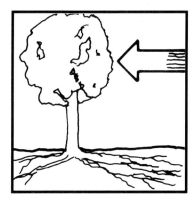

Where there is a strong prevailing wind, main root growth is likely to be on the windward side.

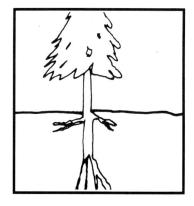

Conifers are likely to have a deep central tap root and few other large roots.

Further Reading:

Building Research Establishment (BRE) Digest 96: "Foundations on shrinkable clay: avoiding damage due to trees".

Building Research Establishment (BRE) Digest 298: " The influence of trees on house foundations in clay soils".

BS 5837: "Code of practice for trees in relation to construction".

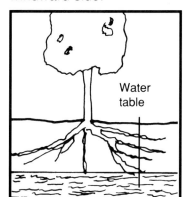

Roots will not extend below the water table.

WHERE TREES OCCUR ON SITE

Note:
Generally the retention of trees on a site is desirable from an amenity point of view and should be considered a worthwhile objective, provided that it does not pose a risk to the fabric and services of the houses on the site.
It should also be noted that the removal of trees and roots may cause the ground to expand after removal.

Rule of thumb:
It may be possible to remove up to 25% of root growth. Consider, however, in what direction new roots may grow in compensation.

Note:
Specialist advice should be obtained where cutting roots is necessary.

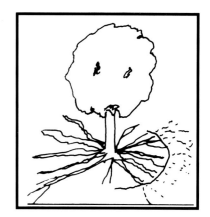

Potential problems and solutions

Problem ▼

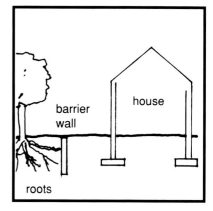

1 Tree root displacing structure.

Solution ▼

2 Cut back root growth, provide barrier wall of 150mm concrete or equivalent.

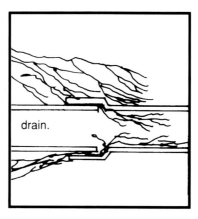

1 Tree root entering drain.

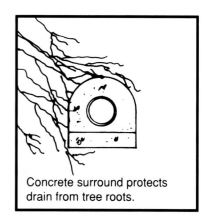

Concrete surround protects drain from tree roots.

2 Encase drain.

INTRODUCTION

Foundations are the most important structural element of any building and great care should be taken to ensure they are constructed properly. The guidance on the following pages should be followed to help ensure this.

Before digging make sure that the location of underground pipes and services is first established.

Remove all vegetable soil from the entire area of the dwelling prior to digging foundations. All organic matter, including roots and timber must be removed. Excavation should be taken down to a firm bottom prior to filling.
Note: Remove soft spots and all loose material prior to pouring of foundation. Backfill with lean mix concrete to make up levels prior to pouring foundation concrete.

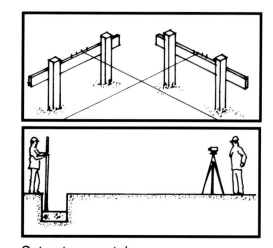

Set out accurately.

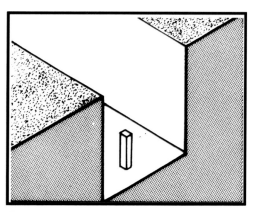

Ensure clean, dry, level, compact bottom, especially in wet weather. Provide pegs for levelling top of foundation.

REINFORCEMENT

If ground conditions require designed reinforced strip foundations, it is essential that steel is placed in accordance with design details and specification and where foundations are stepped that reinforcement is designed and placed to take such steps into account.

In some circumstances, nominal steel reinforcement is introduced in plain concrete foundations. Where this is done, care should be taken that a minimum cover to reinforcement of 75mm is maintained where concrete is poured against an earth surface.

4

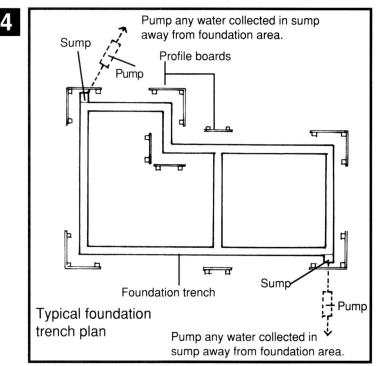

Typical foundation trench plan

Allow ground water to discharge to sump.

5 Where excavations become waterlogged prior to pouring concrete, provide a sump, and clean the trench bottom down to a solid bearing.

6

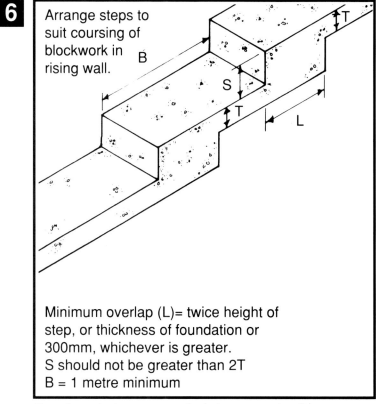

Arrange steps to suit coursing of blockwork in rising wall.

Minimum overlap (L)= twice height of step, or thickness of foundation or 300mm, whichever is greater.
S should not be greater than 2T
B = 1 metre minimum

Step foundations correctly (where required).

TIMBERING TO TRENCHES

Before digging make sure that the location of underground pipes and services is first established.

As a general rule, any excavation in excess of 1250mm must be supported or have the trench sides sloped ("battered") to a safe angle – for deeper trenches the sides may also have to be "benched". Care should also be taken to avoid injury to workers and other persons arising from:

◆ Material falling into the trench

◆ Persons falling into unprotected trenches

◆ Unsafe access – use ladders correctly

◆ Vehicles driving into excavations

◆ Fumes (such as from diesel or petrol engines) entering the excavation.

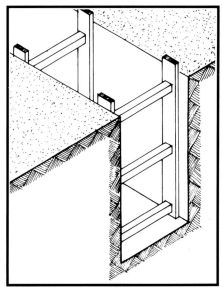

Timbering in compact/firm soils.

Timbering in moderately firm/soft soils.

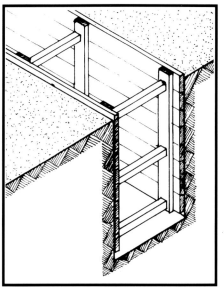

Timbering in dry granular soils/sand.

SLUMP TEST: For concrete

The slump test is used to determine the workability of fresh concrete.

The measurement of workability of fresh concrete is of importance in assessing:

1 The practicability of placing and compacting the mix and

2 Maintaining consistency throughout the job.

In addition, workability tests can be used as an indirect check on the water / cement ratio of the concrete.

The slump test is suitable for normal cohesive concrete mixes of medium to high workability and is the workability test most commonly used. Changes in the value of slump may indicate changes in materials, in the water content or in the proportions of the mix, so it is useful in controlling the quality of the concrete as produced.

The apparatus required for this test consists of:

♦ A truncated conical mould 100mm diameter at the top, 200mm at the bottom and 300mm high.

♦ A steel tamping rod of 16mm diameter and 600mm long with both ends hemispherical.

♦ A measuring rule.

♦ A suitable smooth, rigid and impervious surface.

Test procedure

♦ Ensure the inside of the mould is clean and damp before each test.

♦ Place the mould on a smooth, horizontal and rigid impervious surface.

♦ Hold the mould down using the foot rests and fill with concrete in three layers of approximately equal depth. Tamp each layer with 25 strokes of the tamping rod, the strokes being uniformly distributed over the cross-section of the layer. Each layer should be tamped to its full depth, allowing the rod

to penetrate through into the layer below. The concrete should be heaped above the mould before the top layer is tamped.

♦ When the top layer has been tamped strike off the excess concrete level with the top of the mould by a sawing and rolling motion of the tamping rod. Clean any spillage away from around the base of the mould.

♦ Slowly lift the mould vertically from the concrete.

♦ Turn the mould upside down and place it beside the slumped concrete ensuring it is sitting level. Lay the tamping rod across the base of the upturned mould and using the rule measure the slump from the top of the mould to the highest point on the concrete being tested.

♦ If the concrete collapses or shears off laterally, the test should be repeated with another sample of the same concrete.

♦ If after the slump measurement has been completed the side of the concrete is tapped gently with the tamping rod, a well proportioned cohesive mix will gradually slump further, but a harsh mix is likely to shear or collapse.

Slump testing of concrete.

Slump results:

Low workability = 0-50mm slump.
Medium workability = 50-75mm slump.

TYPICAL CONCRETE MIXES FOR HOUSING:

The table opposite gives examples of recommended concrete mixes for various aspects of housing construction including foundations.

Application	Strength (N/mm^2)	Minimum Cement Content (kg/m^3)	Maximum Water: Cement Ratio	Recommended[3] workability
Lean mix	7.5	100	N/A	Low
Pipe and Kerb bedding	10	160	N/A	Low to Medium
Blinding and sub-floors	10	180	N/A	Low to Medium
Strip Foundations	15	200	0.85	Medium
Ground supported floors	20	220	0.80	Medium
Paths around houses	25	240	0.75	Medium
Reinforced Concrete	Follow designer's specification.			

Notes:

1   Concrete strengths of 25N and greater may be reduced by 5N provided that the cement content and water cement ratio requirements of the table are met.

2   The minimum cement contents shown assume a maximum aggregate size of 20mm.

3   Low workability = 0-50mm slump.
    Medium workability = 50-75mm slump.

4   Heavily trafficked garages will require 25N concrete with minimum cement content of 240kg/m^3 and water/cement ratio of 0.75 and medium workability.

5   Where sulphates are present or where ground conditions might give rise to damage to concrete, the concrete specifications for foundations may require modifications. (See pages 6 and 7)

STRIP FOUNDATIONS

Cracking due to foundation failure.

Wall not centred on foundation.

**STRIP FOUNDATIONS:
HomeBond RULES.**

**PROVIDE
FOUNDATIONS TO ALL
BLOCK AND BRICK
WALLS, CHIMNEYS AND
TO LOAD BEARING
PARTITIONS.
INCLUDING LOAD
BEARING STUD WALLS.**

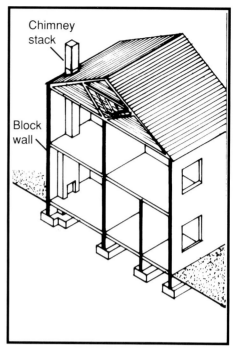

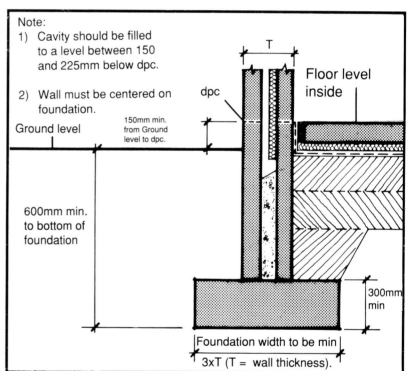

Note:
1) Cavity should be filled to a level between 150 and 225mm below dpc.

2) Wall must be centered on foundation.

Ground level

150mm min. from Ground level to dpc.

600mm min. to bottom of foundation

dpc

T

Floor level inside

300mm min

Foundation width to be min 3xT (T = wall thickness).

Foundation dimensions required by HomeBond:

Depth: Minimum depth of foundation excavation below finished ground level to be **600mm.**

Width: Minimum width of foundation to be three times the thickness of wall it is supporting.

Thickness: Minimum concrete thickness to be **300mm.**

Note: 1 These are minimum figures. Site conditions may dictate greater dimensions or special foundation design such as rafts, piles or foundations on rock.
2 Foundations to large elements such as chimneys and piers require increased foundation sizes.

RAFT FOUNDATIONS

Important:
Raft foundations are a specialist form of construction. It is essential that they be designed and supervised by an appropriately qualified engineer. The engineer appointed must be qualified by examination, be in private practice, and possess professional indemnity insurance.

A site investigation must be carried out by the engineer and the raft designed taking into consideration the result of such investigation. The construction must be supervised and completed to the engineer's satisfaction.

Raft foundation

Hardcore material should be crushed graded stone free from shale, 100mm maximum size.

Demolition material and site rubbish should not be used. Excavated material must not be placed inside the line of perimeter walls.

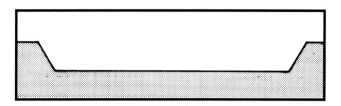

1

Excavate to suitable bearing, ensuring that all soft layers are removed.

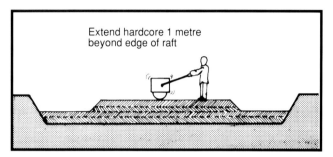

Extend hardcore 1 metre beyond edge of raft

2

Fill and compact in 225mm maximum layers and provide step at edge.
Note: Filling and compaction to be carried out under structural engineer's supervision.

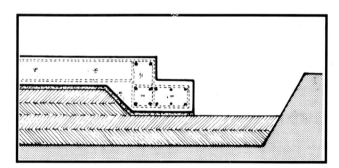

3

Place reinforcement, shutter, pour, vibrate and cure.
Note:
Dimensions/grade of concrete/steel sizes, locations: all to structural engineer's specification.

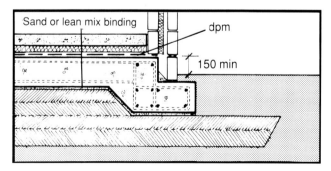

Sand or lean mix binding

dpm

150 min

4

Provide D.P.M. insulation and 65mm thick screed.
Ensure that D.P.M. is 1000 gauge. (i.e. min.0.25mm thick)
Do not use recycled material.
Note:
1. Services such as water/heating pipes must be above the level of the structural slab.
2. Where a screed is placed on top of insulation, screed to be minimum 65 mm thick, and it is recommended that light mesh reinforcement be incorporated.
3. Insulation should not be placed under the raft.

DRAINS NEAR FOUNDATIONS

Drains close to foundations may be subjected to loads from the foundations themselves. Precautions are necessary to avoid settlement of foundations and / or fracturing of drains.

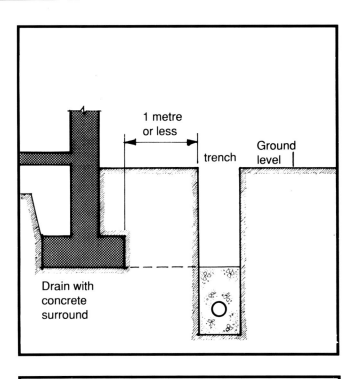

1 metre or less

trench

Ground level

Drain with concrete surround

Where a drain is more than 1 metre from a foundation and at a lower level than the foundation, the drain trench must be filled with concrete to a level at least 150mm above the 45° line of pressure from the foundation.

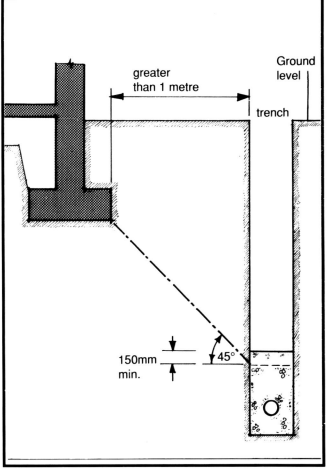

greater than 1 metre

Ground level

trench

150mm min.

45°

SERVICES

Inadequate clearance or excessive rigidity can cause settlement or fracture of service pipes and ducts.
An opening should be formed to provide at least 50mm clearance all around any pipe passing through a rising wall and the opening masked with rigid sheet material to reduce the risk of entry by fill or vermin.

Note:
In the interest of clarity the ope masking has been omitted from the sketches opposite.

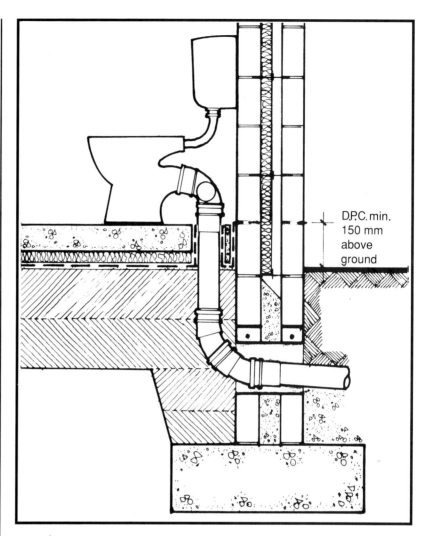

D.P.C. min. 150 mm above ground

Services entering through the rising wall should have 50mm clearance all round pipe to accommodate movement. Ducts should be provided where necessary to facilitate the introduction of services.

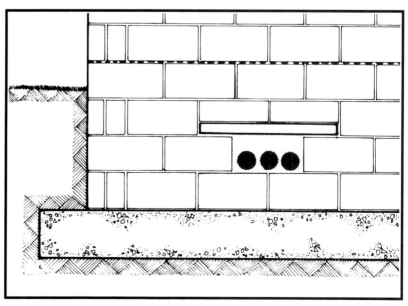

For large openings a lintel should be provided.
Do not use metal lintels in rising walls.

CHANGES OF LEVEL

Where a significant change of level occurs from one side of a rising wall to the other, the wall will, particularly in the partially built state, act as a retaining wall. This necessitates an increase in thickness to resist overturning forces. Great care should be exercised in the compaction of fill material.

The height of the rising wall should not be greater than 3 times the wall thickness. If the rising wall is of cavity construction and the cavity is not filled the thickness is equal to the sum of both leaves as illustrated below.

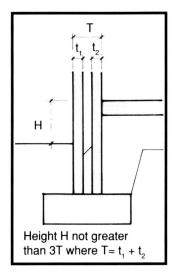

Height H not greater than 3T where $T = t_1 + t_2$

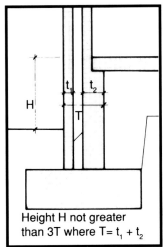

Height H not greater than 3T where $T = t_1 + t_2$

Note: For guidance on appropriate reinforcements and slab depths in suspended reinforced concrete floors see page 40.

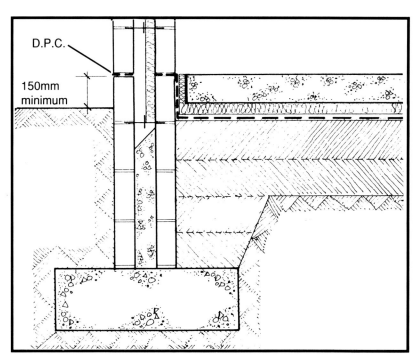

Change of level between 150 and 600mm

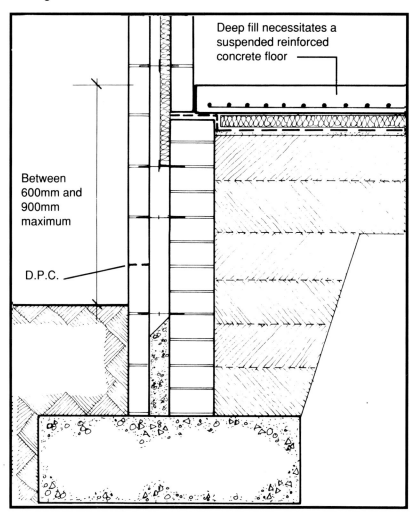

Change of level between 600 and 900mm. Where the change in level exceeds 3 times the wall thickness the wall should be designed by an engineer. The engineer appointed should be qualified by examination, be in private practice and possess professional indemnity insurance.

CHANGES OF LEVEL
continued.

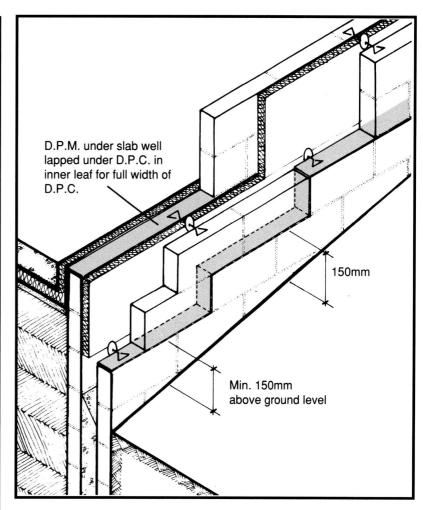

D.P.M. under slab well lapped under D.P.C. in inner leaf for full width of D.P.C.

150mm

Min. 150mm above ground level

Where the site changes level or has a slope, D.P.C. must be stepped with the slope.

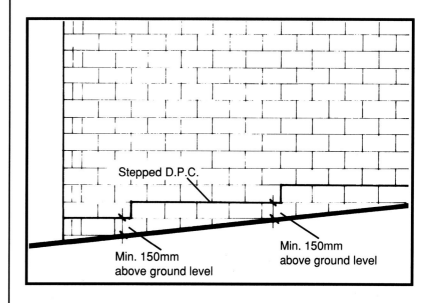

Stepped D.P.C.

Min. 150mm above ground level

Min. 150mm above ground level

CHANGES OF LEVEL
continued.

See Appendix K for underfloor insulation requirements.

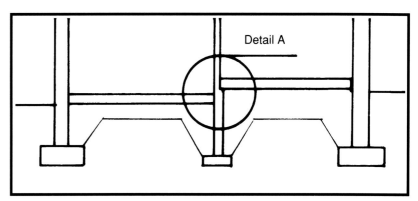

Changes in floor level within the same house

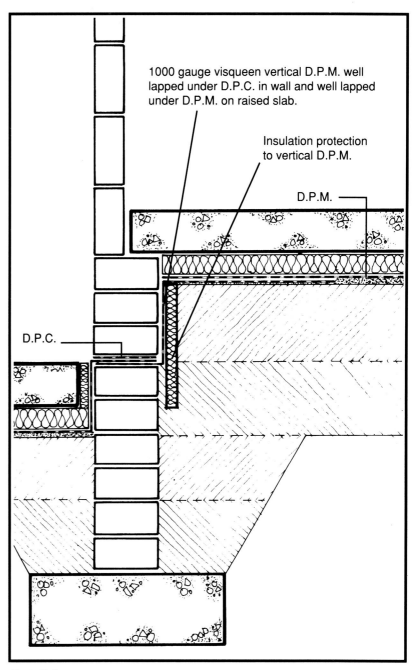

1000 gauge visqueen vertical D.P.M. well lapped under D.P.C. in wall and well lapped under D.P.M. on raised slab.

Insulation protection to vertical D.P.M.

D.P.M.

D.P.C.

Detail A
Vertical joints in the D.P.M. or D.P.C. along the length of the wall should be lapped 150mm min. and taped using a double sided proprietary sealant tape.

CHANGES OF LEVEL
continued.

To avoid the risk of flanking penetration by dampness travelling beneath the higher floor through the party wall above floor level in the lower house it is advisable to construct the party wall in such locations as a cavity wall as illustrated here.

The cavity must be drained to the outside face of the building.

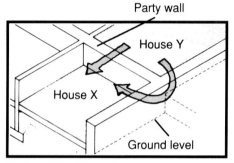

Path of flanking dampness

Note: Where the change in level on opposite sides of the party wall exceeds 600mm see guidelines on page 22 regarding the need to increase wall thickness.

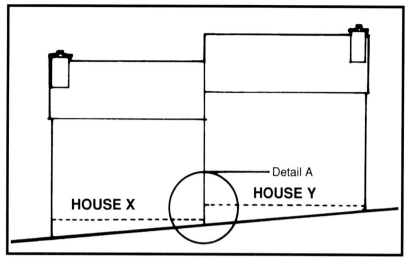

Changes in level between adjoining houses.

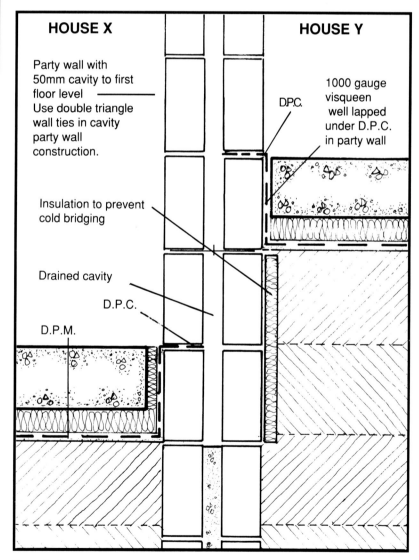

Detail A

CHANGES OF LEVEL
continued.

Where areas of a dwelling are below external ground level great care must be taken to ensure there is no risk of moisture penetration. It is vital in these cases that the architect or engineer designing the house incorporates sufficient barriers to prevent water penetration.

Where a floor level is partially below ground level on a site as illustrated below, it is recommended that the extent of the external wall which is below ground level be built as detailed opposite or alternatively fully tanked if there is a risk of build up of water pressure.

Where an entire floor level is below ground level, e.g. a basement, the external walls below ground level should be tanked as illustrated on pages 27 and 28.

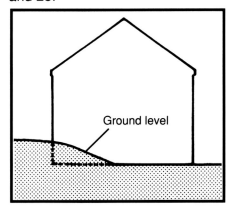

Ground level

Floor level partially below ground level

In this situation there should be no build up of water pressure, and the extent of the cavity below ground level should be drained.

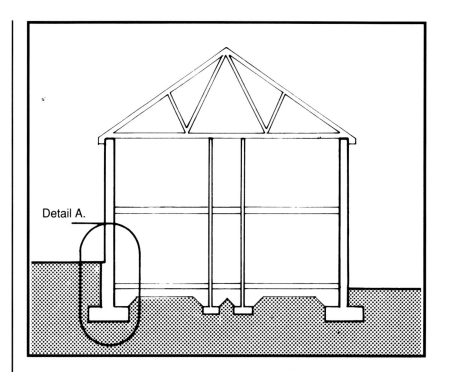

Detail A.

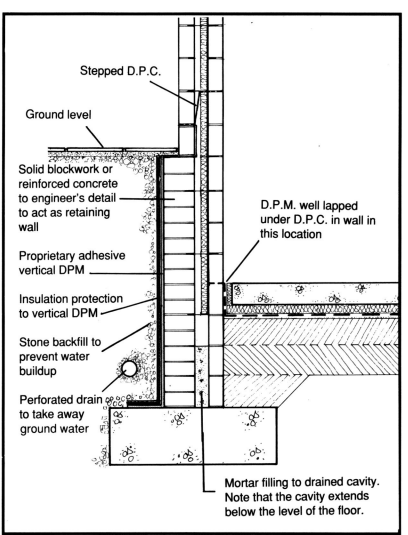

Stepped D.P.C.

Ground level

Solid blockwork or reinforced concrete to engineer's detail to act as retaining wall

D.P.M. well lapped under D.P.C. in wall in this location

Proprietary adhesive vertical DPM

Insulation protection to vertical DPM

Stone backfill to prevent water buildup

Perforated drain to take away ground water

Mortar filling to drained cavity. Note that the cavity extends below the level of the floor.

Detail A
Note: There is merit in wrapping the perforated drain in a geotextile filter cloth to reduce the risk of clogging by migration of fines.

TANKING

Introduction
Where a floor is wholly below ground level, as in the case of basement construction, the entire extent of the external walls below ground level should be tanked.

The principle of tanking is to provide a continuous waterproof membrane of mastic asphalt or prorietary adhesive d.p.m. applied horizontally to the base concrete and in complete continuity with the vertical asphalt or adhesive d.p.m. to the basement walls, externally or internally, according to the nature of the site.

Tanking is necessary as basement and semi-basement constructions are prone to water penetration caused by hydrostatic pressure upon the retaining walls. Additionally contaminants in the ground water can attack and damage in-situ, reinforced concrete, and mass concrete constructions.

Application
The tanking membrane may be sandwiched between two layers.

External tanking: Where the membrane is protected by a protective wall on the excavation side with the structural wall internally, as illustrated opposite.

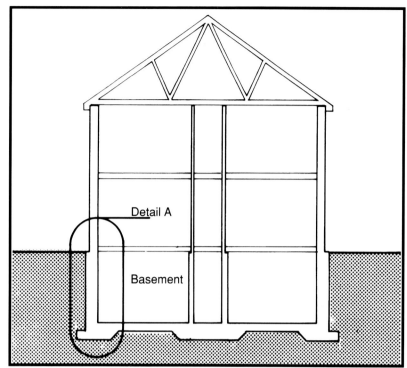

Basement tanking

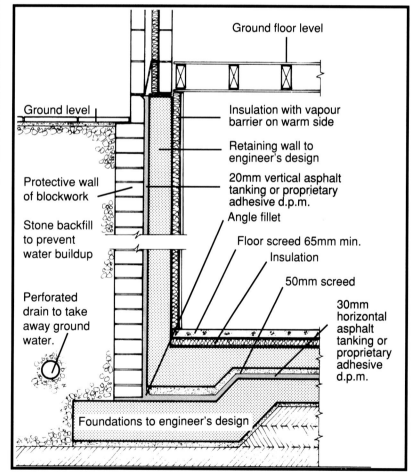

Typical external tanking detail.

Note: If floor screed is laid on top of insulation, screed to be minimum 65mm thick. It is recommended that light reinforcement is incorporated.

Note: There is merit in wrapping the perforated drain in a geotextile filter cloth to reduce the risk of clogging by migration of fines.

TANKING
continued.

Internal tanking: Where the structural wall and floor slab retains the excavation and sandwiches the membrane with an internal skin providing protection.

Of these two methods, external application of mastic asphalt is preferred. It affords protection to the main structure from attack by sulphates in the surrounding soil and ground water.

Thickness and number of coats
On horizontal surfaces and surfaces sloping up to 30° to the horizontal mastic asphalt should be laid in three coats to a total thickness of 30mm.

On vertical surfaces and slopes over 30° to the horizontal mastic asphalt should be applied in three coats to not less than 20mm and taken to a height of at least 150mm above ground level.

Where angle fillets are required they should be at least 50mm wide and applied in two coats at the junction of two planes forming an internal angle.

It is essential the tanking be applied to clean dry surfaces which should be free from sharp protrusions which could puncture the material and destroy the essential waterproofing integrity.

Backfilling
Backfilling to basement walls should be carefully executed in graded material in layers not exceeding 150mm and compacted.

Where pitch fibre or PVC/uPVC drains are used it is important that the granular material be well compacted to the sides of the section to prevent deformation.

Note:
Basement design, construction and tanking is a specialised area and an engineer should be engaged, The engineer appointed should be qualified by examination, be in private practice and possess professional indemnity insurance.

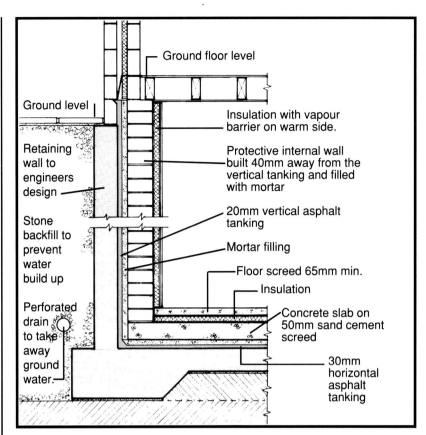

Typical internal tanking detail.

Note: If floor screed is laid on top of insulation, screed to be minimum 65mm thick. It is recommended that light mesh reinforcement is incorporated.

Note: There is merit in wrapping the perforated drain in a geotextile filter cloth to reduce the risk of clogging by migration of fines.

APPLYING AND LAYING TANKING.
◆ **Protection.**
   Provide protection against damage to the membranes at all times. Provide or ensure the provision of the permanent protective construction as soon as practicable after completion of the damp-proofing membrane. Keep the area which has been overlaid clear of materials used by other trades.

◆ **Membrane support.**
   Provide full support without voids over the whole of both surfaces of tanking membranes in a sandwich construction. Where the membrane is applied internally to vertical surfaces, ensure that the protecting inner wall is so constructed as to fully support the membrane.

◆ **Externally applied membrane.**
   Provide and maintain effective temporary protection to the membrane at the junction of floor and wall. Do not drive mechanical fixings through the membrane.

◆ **Temperatures of heated material.**
   Do not heat bonding bitumens above 260°C. Do not heat asphalt above 230°C for prolonged periods. Measure temperatures with thermometers in the heating cauldrons or in the mastic asphalt immediately after it has been removed from a mixer.

INTRODUCTION

Hardcore should be placed and compacted in accordance with the guidance given on these pages. Failure to observe this simple guidance can result in expensive and disruptive remedial work.

Unacceptable fill material.

Failure of fill

1

Hardcore material should be clean graded crushed stone, free from shale, 100mm maximum size.
Demolition material, site rubbish or pit run gravel must not be used.
Excavated material must not be placed inside line of perimeter walls.

Dry rot material in fill

2

Consolidate hardcore in layers not exceeding 225mm in thickness.

This is essential to avoid subsequent settlement of hardcore.
Particular attention should be paid to compacting hardcore where depth increases locally e.g. foundation trenches.

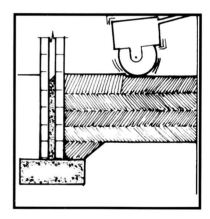

3

Total depth of hardcore should not exceed 900mm, except where a suspended floor is being used.

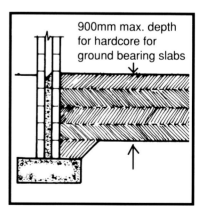

900mm max. depth for hardcore for ground bearing slabs

Where site conditions require a depth of fill of hardcore in excess of 900mm, a suspended floor construction should be used. Suspended floors can be of timber, insitu reinforced concrete or precast concrete.

BLINDING

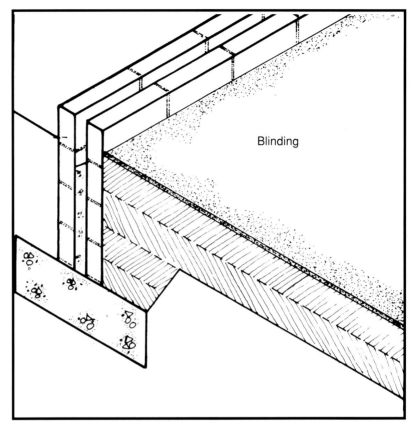

Blinding should be of minimum thickness sufficient only to fill surface voids. For reinforced slabs blinding to be firm and even to support chairs to reinforcement.

Clean hardcore material should be graded crushed stone, free from shale and 100mm maximum size. Demolition material, site rubbish or pit run gravel must not be used. Excavated material must not be placed inside line of perimeter walls.

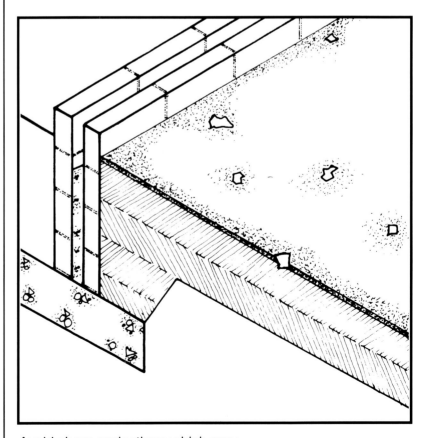

Avoid sharp projections which may puncture the damp proof membrane.

DAMP PROOF MEMBRANE

When purchasing or using D.P.M. the following points should be considered.

1   Only use 1000 gauge material. i.e. 250 microns (0.25mm)

2   Do not use material described as "heavy duty" or "C1000" as such material is not 1000 gauge.

3   Do not use "recycled material". It must be virgin.

4   Avoid sharp projections and damage to the membrane.

5   Take care not to damage the D.P.M. at the junction with the wall D.P.C. when power floating the floor slab.

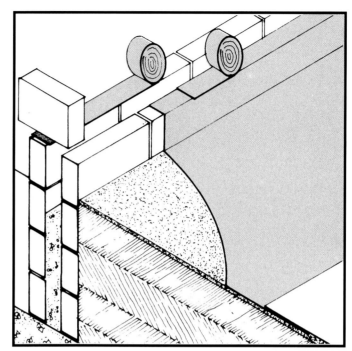

6   D.P.M. should lap under D.P.C. for full thickness of inner leaf.

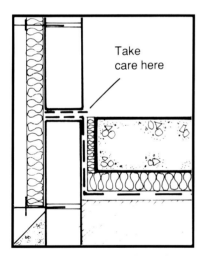

D.P.M. must be in place before walls are built above D.P.C. level so that a proper positive connection between D.P.M. and D.P.C. can be achieved.

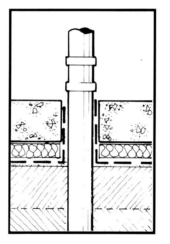

7   Ensure proper seals around services.

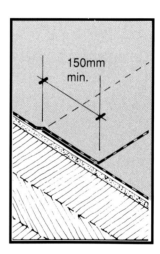

8   Ensure adequate laps.

9   Carelessness causes damage.

UNDERFLOOR VENTILATION

Underfloor ventilation to suspended timber floors is necessary to avoid excessive build-up of moist air in the underfloor space.

Vents should be distributed to ensure thorough ventilation of the underfloor space, to eliminate stagnant air pockets where moisture might accumulate. For detailed guidance see page 33.

Note: Ventilation must be sleeved through cavity to prevent ventilation of cavity.

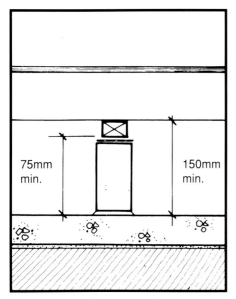

There should be a ventilated air space measuring at least 75mm from the concrete to the underside of the wall plate and at least 150mm to the underside of the suspended timber floor or insulation if provided in this location.

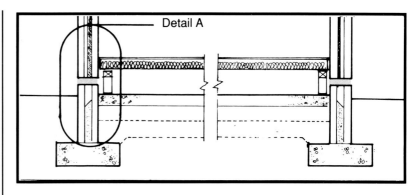

Detail A

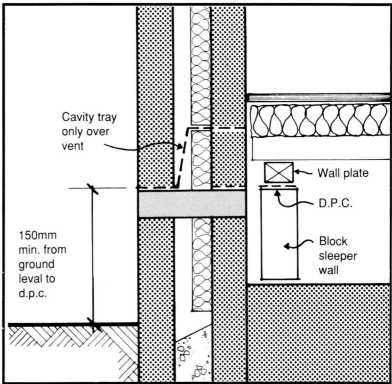

Cavity tray only over vent

150mm min. from ground leval to d.p.c.

Wall plate

D.P.C.

Block sleeper wall

Detail A: Air brick or vent should be sleeved across cavity.

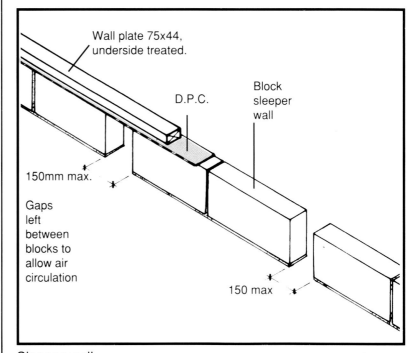

Wall plate 75x44, underside treated.

D.P.C.

Block sleeper wall

150mm max.

Gaps left between blocks to allow air circulation

150 max

Sleeper wall

UNDER FLOOR VENTILATION
continued.

Each external wall must have ventilation openings placed so that ventilating air will have a free path between opposite sides and to all parts of the underfloor void.

The openings should be large enough to give an actual opening at least equivalent to $1500mm^2$ for each metre run of wall.

Any pipes needed to carry ventilating air should have a diameter of 150mm minimum and account should be taken of the fact where such a pipe is used the available free area of the vent is reduced to the area within the circumference of the pipe.

If a 100mm diameter pipe is used, the available free area of the vent will be reduced significantly and a greater number of vents required.

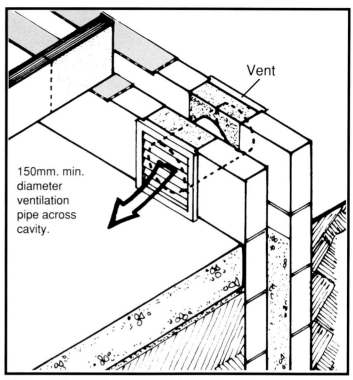

Vent

150mm. min. diameter ventilation pipe across cavity.

Air vent sleeved across cavity

Typical wall ventilators.

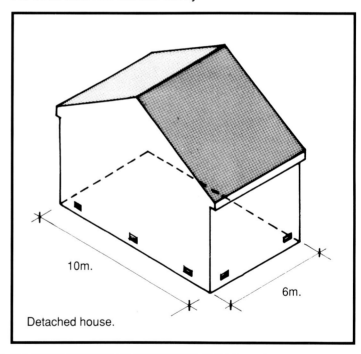

10m.

6m.

Detached house.

How to calculate the number of ventilation opes required, for the house illustrated above.

Procedure:

| Total length of perimeter wall | **X** | Minimum ventilated opening required per meter run of wall = $1500mm^2$ | **÷** | Free air opening of the ventilator being used* | **=** | Total number of ventilated opes required. |

Calculation:

$(32m) \times (1500mm^2) \div (5400mm^2)* = 9$ no. vents min.

For the purpose of balancing air flow use 10 no. vents 2 in the front, 2 in the back, and 3 in each gable.

* This figure will vary depending on the design of the vent and from manufacturer to manufacturer, and on how the vent is sleeved. In this example the figure for a typical purpose designed telescopic wall ventilator is used, i.e., $5400mm^2$.

SUSPENDED TIMBER FLOOR WITH TASSEL WALL

1

Suspended timber floor with tassel walls using standard 115mm x 35mm floor joists of strength class A at 400mm centres, max. span 2.0m. Joist sizes from S.R.11, see pages 114 to 117.

Note: For insulation to suspended timber floors see page 210.

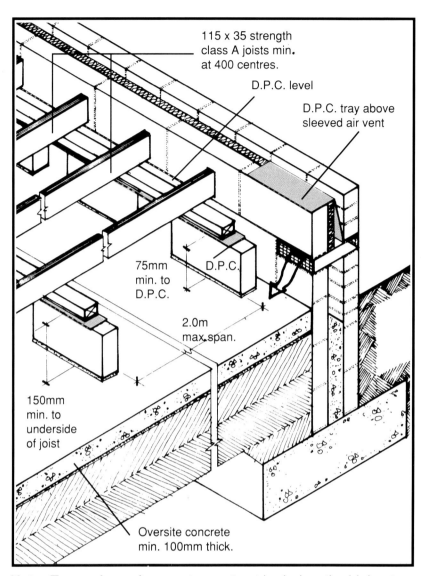

115 x 35 strength class A joists min. at 400 centres.

D.P.C. level

D.P.C. tray above sleeved air vent

75mm min. to D.P.C.

D.P.C.

2.0m max. span.

150mm min. to underside of joist

Oversite concrete min. 100mm thick.

Note: Top surface of concrete must not be below the highest level of the surface of the ground or paving adjacent to the external wall to avoid 'sump' effect.

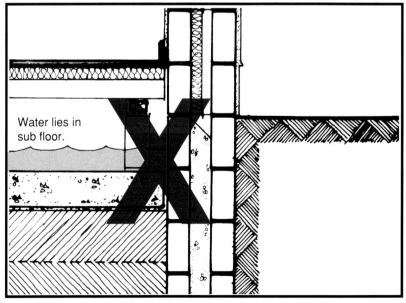

Water lies in sub floor.

Sump effect when subfloor is below ground level
NEVER ALLOW THIS TO HAPPEN

SUSPENDED TIMBER FLOOR WITHOUT TASSEL WALL

2

Suspended timber floors without tassel walls. The joists will typically span from external wall to internal block partition.

Note: For insulation to suspended timber floor see page 210.

Note: In dwellings where there is a mixture of solid floors to some rooms while others have suspended timber floors, it may be necessary to lay ventilating pipes below the concrete subfloor connected to the ventilation openings in the external walls to facilitate through ventilation of the suspended timber floors.

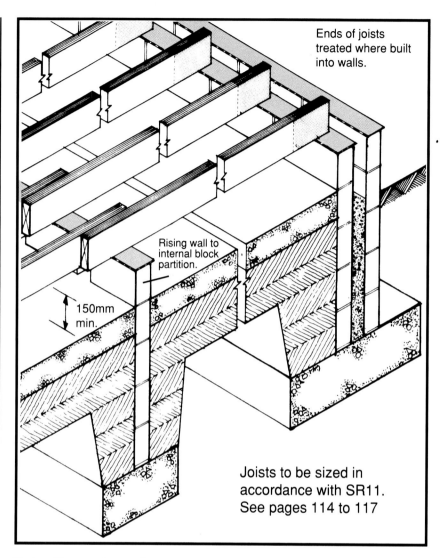

Ends of joists treated where built into walls.

Rising wall to internal block partition.

150mm min.

Joists to be sized in accordance with SR11. See pages 114 to 117

Note: Top surface of concrete subfloor must not be below the highest level of the surface of the ground or paving adjacent to the external wall to avoid 'sump' effect.

1 Solid floating concrete ground floor slab

To be used where the depth of fill does not exceed 900mm.

Note:
For definition of fill material see page 29.

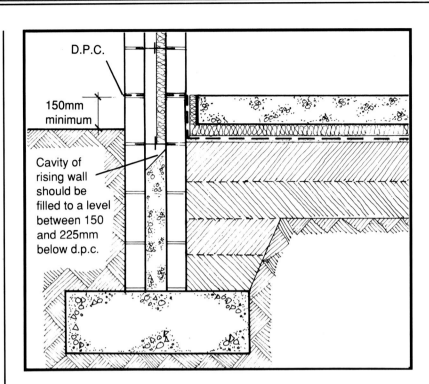

D.P.C.

150mm minimum

Cavity of rising wall should be filled to a level between 150 and 225mm below d.p.c.

2 Cast in-situ suspended ground floor slab

To be used where the depth of fill exceeds 900mm.

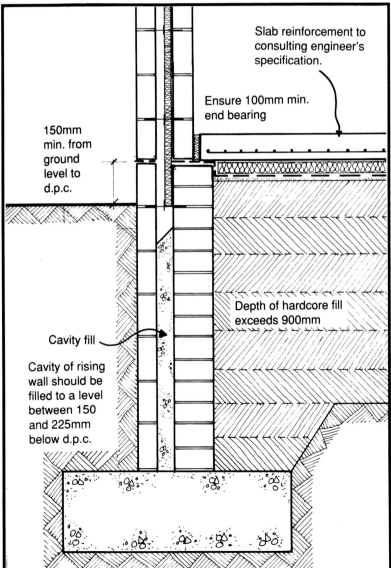

Slab reinforcement to consulting engineer's specification.

Ensure 100mm min. end bearing

150mm min. from ground level to d.p.c.

Depth of hardcore fill exceeds 900mm

Cavity fill

Cavity of rising wall should be filled to a level between 150 and 225mm below d.p.c.

Cast in-situ suspended floor slab external wall detail, where depth of hardcore fill exceeds 900mm.

CAST IN-SITU SUSPENDED FLOOR SLAB PARTY WALL DETAIL

Where a suspended floor slab is continuous over a wall as shown opposite additional reinforcement should be placed in the top of the slab to reduce the risk of cracking.

Explosive gas mixtures in underfloor voids:

There is a risk of explosive gas mixtures accumulating in underfloor voids.

This may occur where:

(1) Building takes place on reclaimed/contaminated land, or
(2) There are gas pipes present and
floor slabs are built clear of the ground or where fill below a suspended ground floor might settle, thereby forming an underfloor void.

In such cases a ventilated air space should be provided measuring at least 150mm clear from the ground to the underside of the floor or insulation (if provided in this location).

Perimeter ventilation must be used in conjunction with the above and in accordance with the guidance given on page 33 to facilitate thorough ventilation.

See also guidance on pages 6 and 7 regarding chemical hazards and contaminants.

Top of floor level must not be below ground level outside to prevent sump situation occurring.

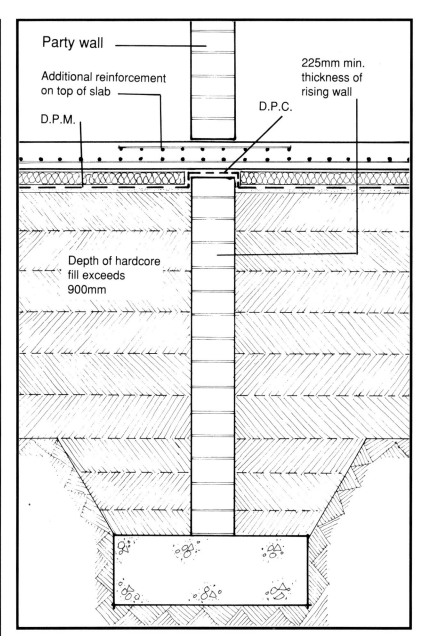

Cast in-situ suspended floor slab party wall detail. Where depth of hardcore fill exceeds 900mm.

PROBLEMS WITH CONCRETE GROUND FLOORS

Introduction.

HomeBond has identified failure in concrete floors as quite a common problem. This failure can be attributed to one or more of the following; poor materials, poor workmanship or bad design.

Care taken at each of these stages can prevent disruptive and expensive repairs at a later date.

These pages outline some of the more common problems and suggest practical solutions to each.

1 **Strip the existing ground down to suitable bearing.**

It is vital that all top soil and vegetable soil is stripped over the entire area of the dwelling before work starts. Leaving mounds of soft earth between the lines of foundation trenches is not appropriate as there is a risk it will not be taken out when hardcore filling starts. Hardcore must be placed on good bearing ground.

2 **Use good quality hardcore - clean well graded broken stone.**

Hardcore must be described as 'hard' material. HomeBond will not accept quarry spoil, pit run gravel, demolition rubble or materials that contain builder's sand, gravel, clay or mud. There must never be any timber or vegetable matter in hardcore.

3 **Compact hardcore in layers with a vibro roller.**

Hardcore must be placed in layers, each not exceeding 225mm in depth. As each layer is placed, it must be well compacted with a vibro roller or whacker. Care taken at this stage will reduce the risk of failure.

4 **Do not use hardcore if the depth of fill is more than 900mm.**

Where the depth of fill exceeds 900mm a suspended concrete floor should be used (see page 36 for guidance) as it is less liable to failure. It is also less expensive.

5 **Blind hardcore.**

Hardcore which has been well compacted in layers should not need much blinding.

Blinding should be quarry dust or sand, spread to ensure that sharp points or edges not put down by the vibro roller are covered. Blinding gives an appropriate surface for the D.P.M. to be laid on.

6 **Use 1000 gauge polythene as D.P.M..**

The damp proof membrane has to stop rising damp and it must not be damaged by traffic when the slab is being poured. Make sure that genuine 1000 gauge polythene comes from the supplier. Do not use material described as 'heavy duty' or 'C1000' and do not use recycled material.

7 **Make sure D.P.C. and D.P.M. lap.**

The D.P.C. and the D.P.M. must form a continuous barrier against moisture penetration from the ground.

HomeBond have noted with concern that a number of builders place the polythene D.P.M. and pour the floor slab only after they have completec-the building of the walls. **HomeBond does not approve this sequence of construction and it must not be used.** The D.P.M. and floor slab must be laid when the walls are at D.P.C. level to ensure a proper lap between D.P.C. and D.P.M. as illustrated.

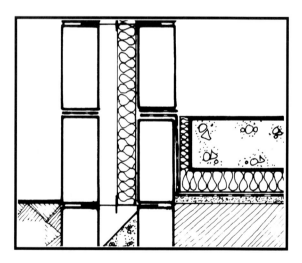

8 **Do not build off the floor slab.**

Floor slabs are generally not constructed to carry loads from walls, piers, chimney breasts and load bearing stud partitions. All blockwork or load bearing stud partitions and rising walls must have foundations. See also page 18.

PROBLEMS WITH CONCRETE GROUND FLOORS
continued.

9 Take care with insulation.
To comply with the Building Regulations it is likely that 55mm thick polystyrene insulation will be required under the entire area of the floor slab. Great care should be taken to ensure this insulation stays in place as the concrete slab is being poured. It may be necessary to weigh down the insulation with bricks, which are then removed as the concrete is poured near to them. To prevent the poured concrete from going under the insulation batts, the batts could be covered with 500 gauge visqueen, or suitably held down.

10 Take care to avoid cracking.
When laying concrete floor slabs and sand cement screeds in hot weather it is necessary to protect them from direct sunlight and keep them cool to prevent drying out too quickly which will result in cracking. Detailed guidance on screeds can be found on pages 216 and 217.

RULES TO PREVENT FAILURE OF FLOOR SLABS

◆ Strip ground of all vegetable matter and top soil.

◆ Use well graded clean broken stone.

◆ Compact hardcore in layers.

◆ Do not use deep fill.

◆ Blind with quarry dust or sand.

◆ Use 1000 Gauge D.P.M.

◆ Place D.P.M. and pour slab when walls are at D.P.C. height.

◆ Never build walls off the slab – they must have foundations.

◆ Floors need insulation to comply with the Building Regulations.

◆ For rafts, D.P.M. and screed must be on top of raft and screed must be at least 65mm thick. It is recommended that light mesh reinforcement is incorporated.

REINFORCEMENT TO SUSPENDED FLOOR SLABS

Reinforcement mesh is available in a range of standard formats. Square mesh with bars at 200mm centres carries the prefix A – the number following the letter A indicates the cross–sectional area of the steel per metre run of mesh, e.g., A393 mesh is a square mesh with 10mm diameter bars at 200mm centres. The standard range of type A meshes is A98 to A393.

Type B meshes have a 200 x 100mm grid with the main steel at 100mm centres and the cross reinforcement at 200mm centres. The range of type B meshes is B196 to B1131.

Type C meshes have a 400 x 100mm grid with the main steel at 100mm centres and the cross reinforcement at 400mm centres. The range of type C meshes is C283 to C785.

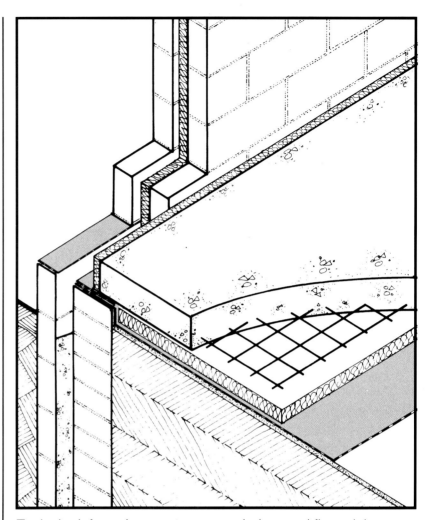

Typical reinforced concrete suspended ground floor slab.

Tabulated below are the appropriate slab depths and reinforcement mesh types for typical domestic ground floor spans. Note that the information given assumes that no internal partitions bear on the slab. The main bars in the mesh should run in the direction of the span and the mesh should be placed so that the main bars are below the secondary bars. There should be a minimum 25mm nominal cover of concrete between the underside of the slab and the main bars. Concrete mix should be 30N20. Slab bearing on supporting rising walls should be at least 100mm.

Max. slab clear span (m)	Slab depth (mm)	Reinforcement
3.0	150	B283 mesh
3.3	150	B385 mesh
4.0	175	B503 mesh
4.4	200	B503 mesh

REINFORCEMENT TO SUSPENDED FLOOR SLABS
continued.

An alternative bearing detail for suspended floor slabs is shown opposite, and to reduce the risk of cracking, additional reinforcement may be placed in the top of slab at the leading edges and particularly at corners.

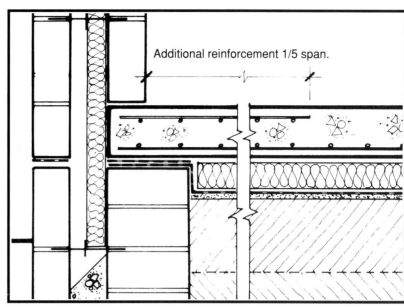

Edge reinforcement.

Where a slab is continuous over a wall, as shown opposite, additional reinforcement should be placed in the top of the slab to reduce the risk of cracking.

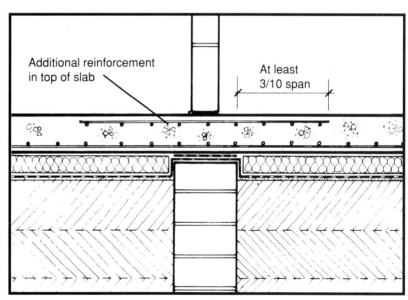

Slab continuous over a wall.

Pipe recesses may only be formed in the perimeter of slabs.

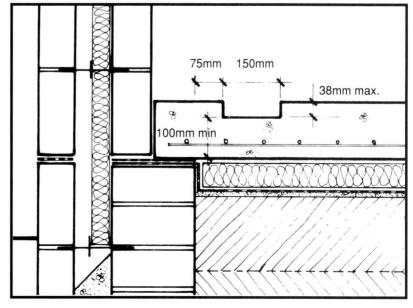

Pipe recesses.

PRECAST CONCRETE FLOORS

As an alternative to the use of suspended timber or suspended in-situ concrete floor slabs, there are available a variety of types of precast concrete floor systems suitable for use in suspended floor construction at ground or upper storey level.

Such precast concrete floors fall into three broad categories as follows:

(1) Hollow slab.

(2) Precast plank or plate with in-situ concrete topping. These types may or may not include infill blocks or void formers.

(3) Beam and block.

This and the following page illustrate examples of these floor types. These illustrations do not represent the products of any particular manufacturer but are meant to be indicative of the range available.

Where systems of the type illustrated are being used, the detailed recommendations of the manufacturer should be followed with particular attention being paid to the following aspects of the installation:

◆ Span and supports. (Some systems will require propping)

◆ Grouting and/or screeding requirements.

◆ Treatment of point loads such as partitions.

◆ Location of damp proof course.

◆ Location, amount and method of installation of insulation to meet the requirements of the Building Regulations and to ensure that cold bridging is minimised.

Typical hollow slab – prestressed – cranage required. Spans of up to 12m readily achievable.

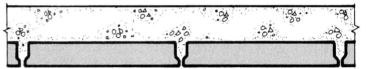

Precast plate with in-situ concrete topping. This method requires propping during construction.

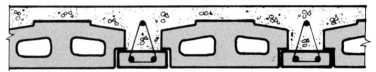

Precast plank with in-situ concrete topping and infill block.

Precast plate with in-situ concrete topping and void formers (shown dotted). Requires cranage during construction.

FLOOR TYPES
continued.

◆ A ventilated air space should be provided measuring at least 150mm clear from the ground to the underside of the floor (or insulation if provided).

See page 33 for underfloor ventilation requirements.

◆ HomeBond ADVISES THAT THE LEVEL OF THE UNDERFLOOR SHOULD BE LEVEL WITH OR ABOVE EXTERNAL GROUND LEVEL TO PREVENT A SUMP SITUATION ARISING.
See page 34.

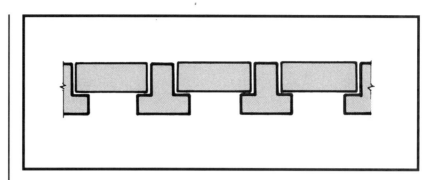

Inverted T-beams with standard concrete blocks.

Note: Beam and block floors are all finished by the application of a concrete screed.

FROM D.P.C. TO ROOF LEVEL

INTRODUCTION

The function of a lintel is to span an opening in a wall and carry the masonry, then transmit this load from the masonry to the wall on either side. Masonry must not be supported on window or door frames which are not designed for the purpose.

Narrow piers between openings give rise to high stress in masonry and the width of the piers must be controlled. For guidance refer to Technical Guidance Document A of the Building Regulations, and to page 76.

Where a lintel span exceeds 3m the lintel may require calculation by an engineer. The engineer engaged should be qualified by examination, be in private practice and possess professional indemnity insurance.

The two most common types of lintel in current use are:

1   Pressed metal lintels.

2   Precast composite lintels.

Detailed guidance on the construction of composite lintels is given on pages 48 to 55 "construction specification".

Pressed metal lintels.
These should be installed in accordance with the manufacturer's instructions regarding load capacity, end bearing, water ingress (D.P.C.), fire and thermal properties as set down in the appropriate Irish Agrément Board Certificate.

The sketches opposite illustrate typical minimum bearing and D.P.C. requirements.

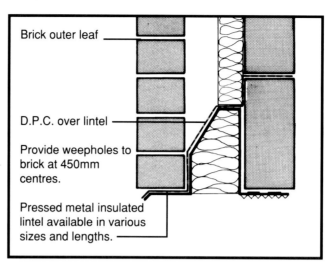

Brick outer leaf

D.P.C. over lintel

Provide weepholes to brick at 450mm centres.

Pressed metal insulated lintel available in various sizes and lengths.

Typical pressed metal lintel

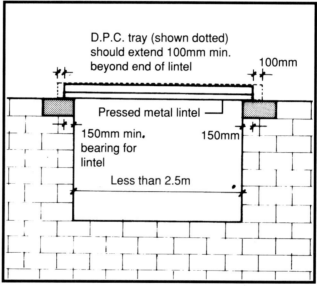

D.P.C. tray (shown dotted) should extend 100mm min. beyond end of lintel

100mm

Pressed metal lintel

150mm min. bearing for lintel

150mm

Less than 2.5m

Lintel bearings: Pressed metal lintel where span is less than 2.5m.

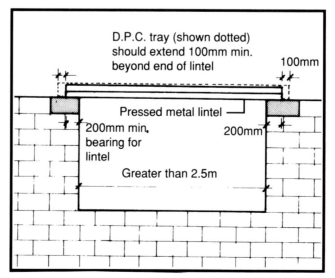

D.P.C. tray (shown dotted) should extend 100mm min. beyond end of lintel

100mm

Pressed metal lintel

200mm min. bearing for lintel

200mm

Greater than 2.5m

Lintel bearings: Pressed metal lintel where span is greater than 2.5m. **Note:** Check manufacturer's requirements for bearing conditions. **Note:** Where outer leaf is brickwork or fairfaced blockwork provide weepholes at 450mm centres (see page 65)

FIRE STOPPING AT DOOR AND WINDOW HEADS IN CAVITY WALL CONSTRUCTION

In semi–detached and terraced houses it is not required to provide fire stopping in the form of a vertical cavity barrier at the junction of party wall and external wall, **provided** that the cavity at door and window heads is closed. This can be achieved using either of the methods illustrated on this page.

Note:

◆ It is not necessary to close the cavity at door and window heads in a detached house. (However, it is likely that conventional construction practice will normally ensure that the cavity is closed at the head).

◆ In all situations the cavity must be closed at wall plate level, and along the top of the gables.

◆ The illustrations in this publication show the semi–detached or terraced situation.

Technical Guidance Document B of the Building Regulations does not make any specific requirements for closing at jambs.

Option

A

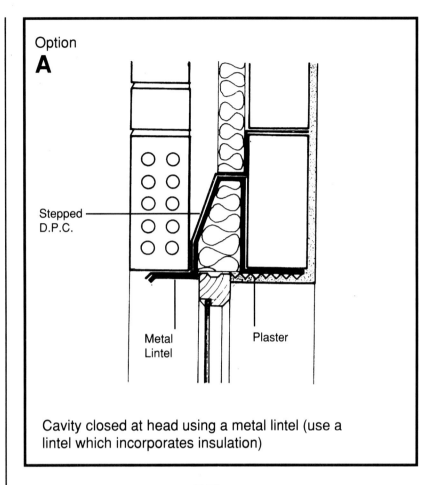

Stepped D.P.C.

Metal Lintel

Plaster

Cavity closed at head using a metal lintel (use a lintel which incorporates insulation)

OR

Option

B

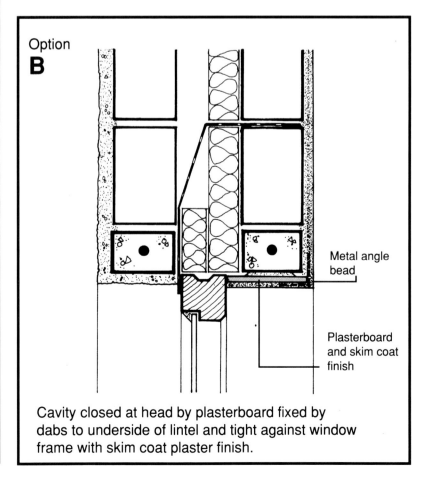

Metal angle bead

Plasterboard and skim coat finish

Cavity closed at head by plasterboard fixed by dabs to underside of lintel and tight against window frame with skim coat plaster finish.

COMPOSITE LINTELS

Introduction
Composite lintels constructed of prestressed concrete lintel acting together with solid block or in-situ concrete should be designed and constructed in accordance with the guidance on the following pages.

The satisfactory performance of a composite lintel depends on the joint action of the prestressed lintel with a zone of blockwork or in-situ concrete laid on top of the lintel — hence the terms 'composite lintel', 'compression blockwork', 'compression zone'.
In the same way that the steel reinforcement of a reinforced concrete beam resists tensile force and the concrete resists the compressive force so in a composite lintel the prestressing tendon in the lintel (i.e. the reinforcing bar) resists the tensile force and the blockwork or in-situ concrete resists the compressive force.

Remember
A prestressed concrete lintel has no strength until combined with the blockwork or in-situ concrete built above it.

PRESTRESSED CONCRETE LINTELS.

Material specification:
Lintels should be manufactured in accordance with I.S. 240: 1980.

Precast prestressed concrete units for use in composite lintels:

Unit dimensions (mm)	Unit Weight (kg/m)	Initial Prestress (kNs)
100 x 65	16	65
150 x 65	24	64
215 x 65	37	130

Concrete strength:
$35N/mm^2$ at transfer of prestress and $40N/mm^2$ at 28 days.

Initial prestress:
56kN (70% ultimate strength of 9.3mm strand).

WALLING MATERIALS:

(a) Concrete masonry

Concrete Blocks
Complying with IS 20: 1987 "Concrete Building Blocks Part 1, Normal Density Blocks".
Solid blocks only to be used in the area of composite action.

Concrete bricks
Complying with IS 189: "1974 Concrete Building Bricks". Bricks of external quality (i.e. minimum $15N/mm^2$) must be used.

Mortar
Complying with IS 406: 1987 "Masonry Mortars".
Mortar designation iii,
1:1:6 Cement : Lime : Sand.
or,
1 : 6 Cement : Sand with plasticiser.'

(b) In-situ concrete.
Complying with IS 326 "Code of Practice for Structural Concrete".
Characteristic strength: $30N/mm^2$.

CONSTRUCTION SPECIFICATION

◆ **Lintel bearing:**
As specified in I.S. 240: min. 150mm for spans up to 1.5m, 200mm for spans from 1.5m to 3.0m.

◆ **Lintel bedding:**
Lintels should be bedded in mortar at supports.
The masonry should be constructed so that lintels bear onto whole solid blocks wherever possible.

◆ **Propping during construction:**
Lintels must be propped at maximum 1.2m centres until composite masonry or concrete has matured.

◆ **Filling of mortar joints:**
Horizontal and vertical joints in the masonry in the composite lintel area should be fully filled. Shell bedding is not allowed in composite lintels.

◆ **Placing of in-situ concrete:**
Watertight shuttering, for the composite lintel should be wetted before placing concrete. The workability of the mix should be such that it can be fully compacted by the vibration techniques available.

◆ **215mm hollow block walls:**
Solid blocks must be used within the area of masonry required for composite action.

◆ **Joist hangers, D.P.C's etc:**
Floor joists, joist hangers, D.P.C. or any other ancillary components must not be allowed to impose a load or interfere with masonry bond within the area of the composite action. D.P.C. must not be built into compression zone.

◆ **Cavity walls:**
Cavity walls should be constructed using a separate lintel under each wall leaf to obviate the cold bridge which a single lintel would create. Additional wall ties increasing the I.S. 325 requirements to 6 ties per square metre (i.e. wall ties at 225mm vertical centres) should be provided within the area of composite action.

◆ **D.P.C.**

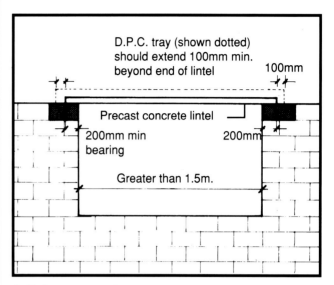

D.P.C. tray (shown dotted) should extend 100mm min. beyond end of lintel   100mm

Precast concrete lintel

200mm min bearing   200mm

Greater than 1.5m.

D.P.C. to extend 100mm min. beyond end of lintel

Warning.
Never cut out or leave out any openings for services in the blockwork course(s) or insitu concrete above a prestressed lintel which acts as the composite masonry.

DESIGN AND LOAD ASSUMPTION

Timber floor loads:
This design method is suitable for domestic dwellings with timber floors up to three storeys high.

Lintels supporting concrete floors are not covered by this guidance.

In drawing up these design tables the following assumptions have been made:

Timber floor loads (kN/m²):

Timber floor loads kN/m²

Self weight of floor:	0.30kN/m²
Imposed load:	1.50kN/m²
Total	1.80kN/m²

Note: the applied load in table 1 below has been derived by multiplying 1.8kN/m² by half the floor span.
e.g. Floor span 6m then the applied floor load on the lintel is (6m÷2) x 1.8kN/m² =5.4kN/m

Table 1. Applied floor load on lintel (kN/m run)	
Span of floor (metres)	Load on lintel kN/m
2m (or less)	1.8
3m	2.7
4m	3.6
5m	4.5
6m	5.4

Pitched roof loads
This design method is suitable for simple plan form roofs of domestic dwellings up to three storeys in height.

Roof construction
Modern roof construction with timber trusses spreads all the roof load to the outside supporting walls and table 2 has been prepared for this situation.

In traditional or cut roof construction some of the load is spread through purlins and struts onto the internal walls. Where this occurs the loads given in table 2 may be reduced by one third to allow for this. Where there are no purlins no reduction may be made.

In assessing dead and imposed loads in roof construction the following assumptions were made:

Dead load on slope:

-Self weight concrete tile roof 0.68 kN/m²
-Self weight fibre cement slate roof   0.25 kN/m²

Dead load on plan:
-Ceiling ties   0.25kN/m²

Imposed load on plan:
-Roof pitch 0° to 30°   0.75 kN/m²
-Roof pitch 30° to 45°   0.75 to 0.0 kN/m²

The increase in dead load of the roof with increasing pitch is sufficiently balanced by the decreasing imposed load to allow the roof loads to be simplified to:

2.0 kN/m² for concrete tiled roofs and
1.6 kN/m² for fibre cement slated roofs
up to 45° pitch.

Note: The lintel applied load in table 2 has been derived by multiplying the appropriate roof load by half the span, e.g. a roof with a 20° pitch and concrete tile finish spans 8m, therefore the applied load on the lintel is 2.0 kN/m² x 4m (half roof span) = 8kN/m (as per table 2 below).

Table 2. Applied roof load on lintel (kN/m) For roofs up to 45° pitch		
Roof span on plan	Concrete tiles	Fibre cement slates
3m	3.0	2.4
4m	4.0	3.2
5m	5.0	4.0
6m	6.0	4.8
7m	7.0	5.6
8m	8.0	6.4
9m	9.0	7.2

COMPOSITE LINTELS OF MASONRY IN CAVITY WALLS

Table 3

Composite depth (D) in mm of 100mm thick block or brick masonry												
Clear span of lintel in metres	Applied load on lintel from Table 1 and/or 2 KiloNewtons per metre (kN/m)											
	1	2	3	4	5	6	7	8	9	10	11	12
0.5	75	75	75	75	75	75	75	75	75	150	150	150
1.0	75	75	75	150	150	150	225	225	225	300	300	300
1.5	75	75	75	150	225	225	300	300	375	375	450	450
2.0	150	150	225	225	300	375	375	450	–	–	–	–
2.5	150	150	300	300	375	450	–	–	–	–	–	–

Note:
1 Only solid concrete blocks or bricks may be used in the area of composite action.
2 Joists, joist hangers, D.P.C.'s or flashings must not be built in to the zone of composite action of the lintel.

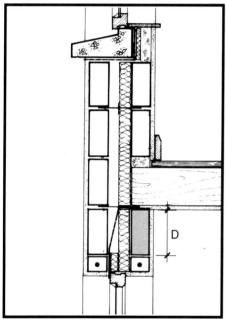

Ground floor window head
Use table 1 to calculate applied floor load on lintel and table 3 to calculate depth of composite blockwork D.

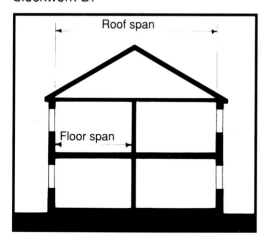

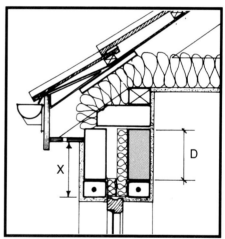

First floor window head
Use table 2 to calculate applied roof load on lintel and table 3 to calculate depth of composite blockwork D.
Note: The solid block cavity closer may be included in the composite depth D.
Note: In exposed conditions or if dimension X exceeds 1m, use a stepped D.P.C. in this location. Such D.P.C. to be used in all brickwork construction.

For example: To calculate the depth of composite blockwork at (1) ground floor window head and (2) first floor window head, for a dwelling with a roof span of 8m, concrete tile roof finish, 30° roof pitch, floor span 4m, cavity wall construction and lintels with a maximum clear span of 2m

(1) Ground floor window head: floor span 4m so from table 1:
Applied floor load on lintel = 3.6kN/m;
Lintel span 2m so from table 3 composite depth (D) = 225mm.

(2) First floor window head: roof span 8m, roof pitch 30°, concrete tile roof finish; so from table 2:
Applied roof load on lintel = 8.0 kN/m
Lintel span 2m, so from table 3, composite depth (D) = 450mm.

COMPOSITE LINTELS OF MASONRY IN 215mm WALLS

Table 4

Composite depth (D) in mm of solid block or brick masonry												
Clear span of lintel in metres	Applied load on lintel from Table 1 and/or 2 KiloNewtons per metre (kN/m)											
	1	2	3	4	5	6	7	8	9	10	11	12
0.5	110	110	110	110	110	110	110	110	110	110	110	110
1.0	110	110	110	110	110	110	110	110	110	110	110	220
1.5	110	110	110	110	110	110	220	220	220	220	220	220
2.0	110	110	110	220	220	220	220	220	330	330	330	330
2.5	110	220	220	220	220	220	330	330	330	330	440	440
3.0	220	220	220	330	330	330	330	440	440	440	–	–
3.5	220	330	330	330	330	440	440	440	–	–	–	–

Note:
1 Only solid concrete blocks or bricks may be used in the area of composite action.
2 Joists, joist hangers, D.P.C.'s or flashings must not be built in to the zone of composite action of the lintel.

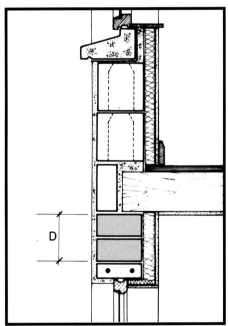

Ground floor window head
Use table 1 to calculate applied floor load on lintel and table 4 to calculate depth of composite blockwork D.

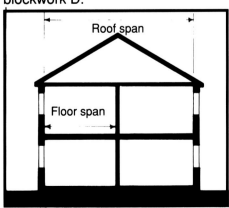

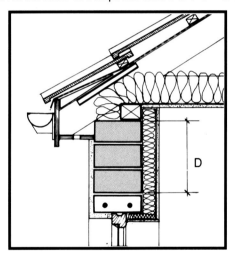

First floor window head
Use table 2 to calculate applied roof load on lintel and table 4 to calculate depth of composite blockwork D.

For example: To calculate the depth of composite blockwork at (1) ground floor window head and (2) first floor window head, for a dwelling with a roof span of 8m, concrete tile roof finish, 25° roof pitch, floor span 4m, hollow block wall construction and lintels with a maximum clear span of 1.5m

(1) Ground floor window head: floor span 4m so from table 1:
Applied floor load on lintel = 3.6kN/m;
Lintel span 1.5m so from table 4 composite depth (D) = 110mm.

(2) First floor window head: roof span 8m, roof pitch 25° concrete tile roof finish so from table 2:
Applied roof load on lintel = 8.0 kN/m;
Lintel span 1.5m, so from table 4;
composite depth (D) = 220mm.

COMPOSITE LINTELS OF IN-SITU CONCRETE IN CAVITY WALLS

Table 5

Composite depth (D) in mm of 100mm thick concrete for lintels in cavity walls												
Clear span of lintel in metres	Applied load on lintel from Table 1 and/or 2 KiloNewtons per metre (kN/m)											
	1	2	3	4	5	6	7	8	9	10	11	12
0.5	75	75	75	75	75	75	75	75	75	75	75	75
1.0	75	75	75	75	75	75	75	75	75	75	75	150
1.5	75	75	75	75	75	75	150	150	150	150	225	225
2.0	75	75	75	75	150	150	225	225	300	300	375	375
2.5	75	75	150	150	225	300	375	375	375	–	–	–

Note:

1. Concrete of 30N/mm² characteristic strength must be used in concrete composite lintels.
2. Joists, joist hangers, D.P.C.'s or flashings must not be built in to the zone of composite action of the lintel.

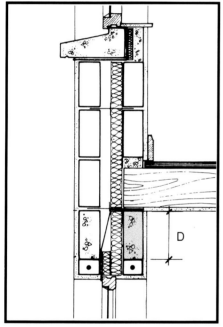

Ground floor window head
Use table 1 to calculate applied floor load on lintel and table 5 to calculate depth of composite in-situ concrete D.

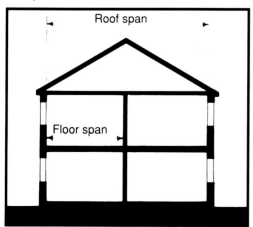

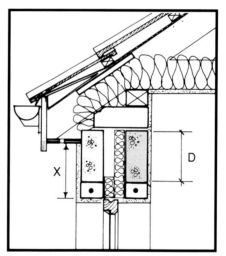

First floor window head
Use table 2 to calculate applied roof load on lintel and table 5 to calculate depth of composite in-situ concrete D.
Note: The solid block cavity closer may be included in the composite depth D.
Note: In exposed conditions or if dimension X exceeds 1m, use a stepped D.P.C. in this location. Such D.P.C. to be used in all brickwork construction.

For example: To calculate the depth of composite in-situ concrete at (1) ground floor window head and (2) first floor window head, for a dwelling with a roof span of 9m, concrete tile roof finish, 20° roof pitch, floor span 4.5m, cavity wall construction and lintels with a maximum clear span of 2m

(1) Ground floor window head: floor span 4.5m so from table 1:
Applied floor load on lintel = 4.5kN/m;
Lintel span 2m so from table 5 composite depth (D) = 150mm.

(2) First floor window head: roof span 9m, roof pitch 20° concrete tile roof finish so from table 2:
Applied roof load on lintel = 9.0 kN/m;
Lintel span 2m, so from table 5;
composite depth (D) = 300mm.

COMPOSITE LINTELS OF IN-SITU CONCRETE IN 215mm WALLS

Table 6

Composite depth (D) in mm of 215mm thick concrete for lintels in 215mm walls												
Clear span of lintel in metres	Applied load on lintel from Table 1 and/or 2 KiloNewtons per metre (kN/m)											
	1	2	3	4	5	6	7	8	9	10	11	12
0.5	75	75	75	75	75	75	75	75	75	75	75	75
1.0	75	75	75	75	75	75	75	75	75	75	75	75
1.5	75	75	75	75	75	75	75	75	75	75	75	75
2.0	75	75	75	75	75	75	75	75	150	150	150	150
2.5	75	75	75	75	75	150	150	150	150	225	300	300
3.0	75	75	75	150	150	150	225	300	300	375	375	375
3.5	75	75	150	150	225	300	375	375	375	–	–	–

Note:

1   Concrete of 30N/mm² characteristic strength must be used in concrete composite lintels.
2   Joists, joist hangers, D.P.C.'s or flashings must not be built in to the zone of composite action of the lintel.

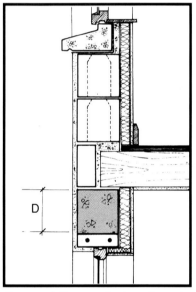

Ground floor window head
Use table 1 to calculate applied floor load on lintel and table 6 to calculate depth of composite in-situ concrete D.

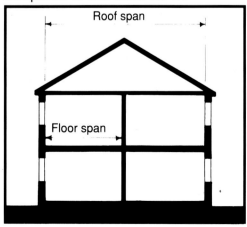

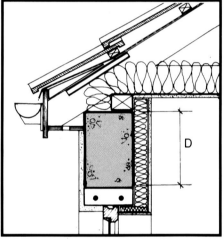

First floor window head
Use table 2 to calculate applied roof load on lintel and table 6 to calculate depth of composite in-situ concrete D.

For example: To calculate the depth of composite in-situ concrete at (1) ground floor window head and (2) first floor window head, for a dwelling with a roof span of 8m, concrete tile roof finish, 30° roof pitch, floor span 4m, hollow block wall construction and lintels with a maximum clear span of 1.5m.

(1)   Ground floor window head: floor span 4m so from table 1:
Applied floor load on lintel = 3.6kN/m;
Lintel span 1.5m so from table 6 composite depth (D) = 75mm.

(2)   First floor window head: roof span 8m, roof pitch 30° concrete tile roof finish so from table 2:
Applied roof load on lintel = 8.0 kN/m;
Lintel span 1.5m, so from table 6;
composite depth (D) = 75mm.

DORMER ROOF CONSTRUCTION

In houses incorporating accommodation in the roof space (i.e. dormer roof) lintels in external walls directly below roof level must carry both roof and floor loads.

The additional load on the lintel from the floor of the dormer accommodation may be calculated from table 1. This should be added to the roof load derived from table 2 and the result used to determine the depth of composite lintel required.

The reduction in load for cut roof construction allowed in table 2 should not be taken.

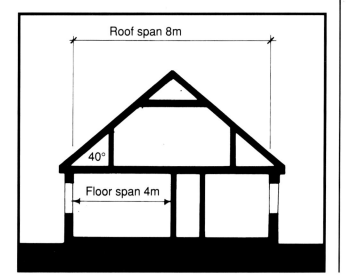

For example:
Calculate the depth of composite blockwork at roof level for a dwelling with a dormer roof spanning 8m, concrete tile roof finish, 40° pitch, floor span of 4m, cavity wall construction and lintels with a maximum clear span of 1.5m.

Stage 1:
Floor span is 4m, so from table 1, the applied floor load on lintel = 3.6kN/m

Stage 2:
Roof span is 8m, roof pitch is 40° with concrete tile roof finish. So from table 2, the applied roof load on lintel = 8.0kN/m.

Stage 3:
Add (applied floor load on lintel) 3.6kN/m + (applied roof load on lintel) 8.0kN/m = 11.6kN/m. Lintel span is 1.5m so from table 3, reproduced below, the depth of composite blockwork D = 450mm.

COMPOSITE LINTELS OF MASONRY IN CAVITY WALLS
Table 3

Composite depth (D) in mm of 100mm thick block or brick masonry												
Clear span of lintel in metres	Applied load on lintel from Table 1 and/or 2 KiloNewtons per metre (kN/m)											
	1	2	3	4	5	6	7	8	9	10	11	12
0.5	75	75	75	75	75	75	75	75	75	150	150	150
1.0	75	75	75	150	150	150	225	225	225	300	300	300
1.5	75	75	75	150	225	225	300	300	375	375	450	450
2.0	150	150	225	225	300	375	375	450	–	–	–	–
2.5	150	150	300	300	375	450	–	–	–	–	–	–

Note:
1 Only solid concrete blocks or bricks may be used in the area of composite action.
2 Joists, joist hangers, D.P.C.'s or flashings must not be built in to the zone of composite action of the lintel.

CAST IN–SITU REINFORCED CONCRETE LINTELS

In any situation where cast in-situ reinforced concrete lintels are being used the guidance regarding ope span, lintel depth and reinforcement given in the table opposite should be used.

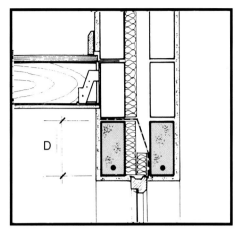

Ope span	Lintel depth(D)	Lintel reinforcement
1.8m	215mm	One T16 bar
2.1m	215mm	One T20 bar
2.4m	300mm	One T20 bar

Note:

1 This guidance should only be used where the roof span is less than 10m. Where the span exceeds 10m an engineer should be engaged. The engineer appointed should be qualified by examination, in private practice and possess professional indemnity insurance.

2	215mm min. lintel bearing.	Cover to reinforcement: 20mm for lintels on internal walls 40mm for lintels on external walls (including lintels on inner leaf of cavity walls)
3	Concrete grade 35N 20.	
4	'T' denotes high yield steel.	

METAL SUPPORT ANGLES TO OUTER LEAF

Where metal angles are used to provide support to the outer leaf the guidance regarding ope span and angle dimensions given in the table opposite should be used.

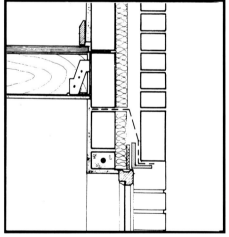

Ope span	Angle required	
	Galvanised mild steel	Stainless steel
1.5m	90 x 90 x 6	90 x 90 x 6
2.0m	*150 x 90 x 10	*100 x 90 x 6
3.0m	*150 x 90 x 10	*150 x 90 x 10

Note:

1 * = Vertical leg of angle   2 200mm min. bearing

3. Galvanised mild steel to BS 729 and 5493

4. Stainless steel to BS 1449

WALL TIES

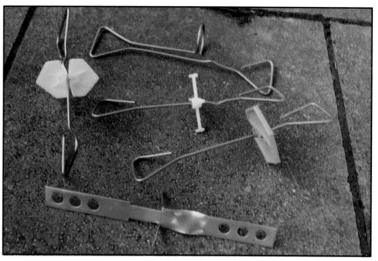

Typical types of wall ties
Wall ties are available in a number of materials and types.

Care should be taken to ensure that:

◆ All wall ties should comply with I.S. 268: 1986: Metal wall ties for masonry walls.
Galvanised wall ties should be galvanised as required by B.S. 1243: 1981: Specification for metal ties for cavity wall construction. Check this with the supplier. Alternatively use a stainless steel wall tie – the use of stainless steel ties is strongly recommended. **Note:** The use of plastic wall ties does not comply with the recommendations of Building Regulations Technical Guidance Document A or the provisions of I.S. 268: 1986.

◆ Wall ties are long enough to ensure 50mm embedment into mortar bed in each leaf, i.e. for a 100mm cavity, use a wall tie at least 200mm long.

◆ Where cavity insulation slabs are used ensure that the tie used suits the insulation and holds it firmly in place. Suitable ties are usually available from the suppliers/manufacturers of wall insulation batts.

◆ Install the ties with the drip downward.

◆ Install the ties as work progresses. Do not push ties into mortar after bricks or blocks are laid.

◆ Ensure there is no backfall (i.e. falling inwards) on wall ties.

◆ Use double triangle wall ties where party walls are of cavity construction; for example, on sloping sites, or at changes of roof level.

◆ Build up both leaves together. No leaf to be more than 1 metre in height above the other.

WALL TIES

Correct spacing and distribution of wall ties is essential for structural stability.
Vertical coursing of wall ties must be staggered.

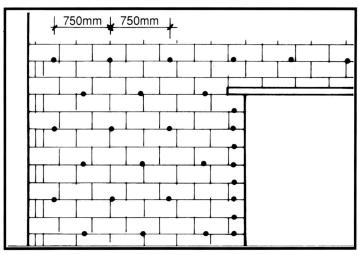

Maximum spacing of wall ties

MAXIMUM SPACING OF WALL TIES		
Cavity width mm	**Horizontal spacing mm**	**Vertical spacing mm**
50 – 75	900	450*
76 –100	750•	450*•

At unbonded jambs to all openings in walls with cavity widths between 50 – 100mm provide wall ties at 225mm vertical centres within 150mm of opening.

* or such spacing as will maintain the same number of ties per square metre.

• Vertical twist ties should be used or ties of equivalent performance.

PROBLEM AREAS

HomeBond have identified a number of recurring problem areas in cavity wall construction associated with damp penetration across the cavity. The seven more common problems are illustrated on the following pages and care should be taken to avoid these.

Keep cavities clean by the use of boards or cavity laths and/or daily cleaning of wall ties.
Provide temporary openings at base of wall.

Board across cavity.

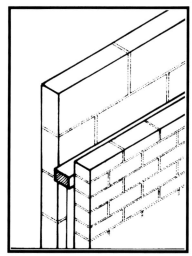

Cavity lath.

Note: All of these recurrent faults can be avoided by care and vigilance on the job.

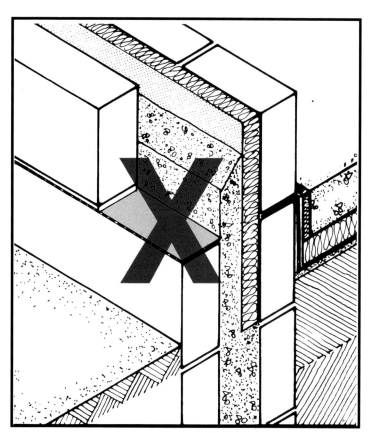

Mortar filling to cavity too high: Build up of mortar droppings. Cavity should be clear for at least 150mm below D.P.C. level.

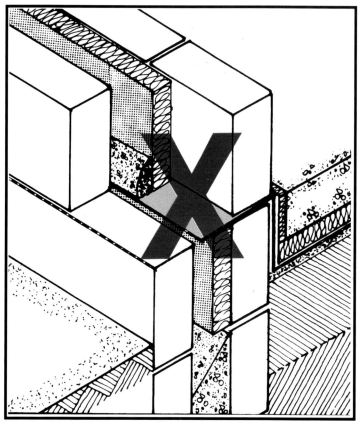

Projecting D.P.C. catches mortar:
Ensure that D.P.C. does not project into cavity at ground floor level.

PROBLEM AREAS
continued.

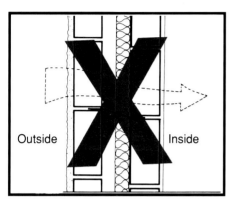

Sloping wall ties act as a bridge across the cavity

3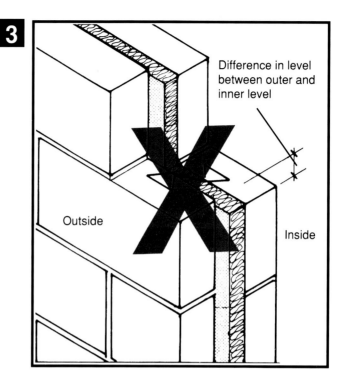

Difference in level between outer and inner level

Outside

Inside

Ties sloping inward act as a bridge across cavity. Take care with coursing of both leaves to avoid backfall on wall ties.

4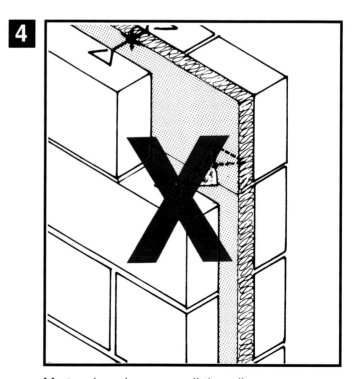

Mortar droppings on wall ties allows penetration of dampness.
Keep cavities and wall ties clean at all times, by using cavity laths or daily cleaning of wall ties as illustrated on page 59.

PROBLEM AREAS
continued.

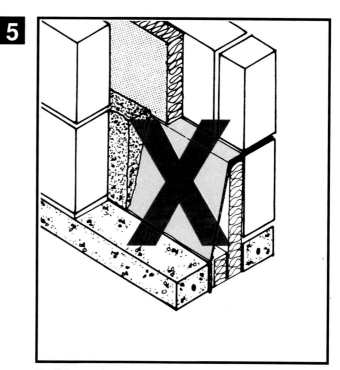

Build up of mortar on trays/lintels over heads –
keep cavities clean as work progresses.

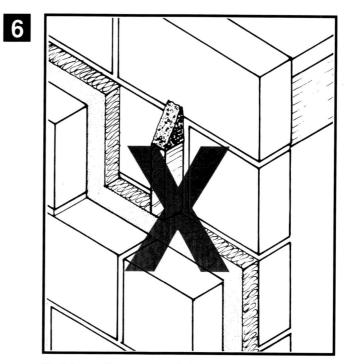

Build up of mortar on joists/purlins projections or similar.
Avoid such projections.
Note: Treat timber ends with coloured preservative.

PROBLEM AREAS
continued.

Note: All of these recurring
faults can be avoided by care
and vigilance on the job.

7

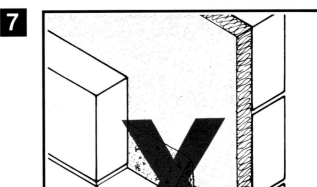

Insulation not restrained against inner leaf.
Use wall ties which are appropriate for the insulation
being used, as per manufacturer's instructions.

DAMP PROOF COURSES

A key element in cavity construction.

The correct use of damp proof courses in cavity wall construction is vital to the satisfactory performance of the building. The following pages highlight some of the principal areas for attention.

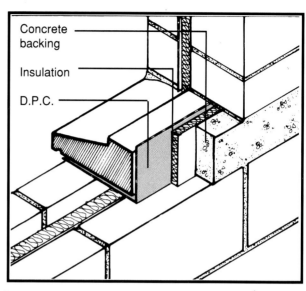

Cill:
Provide D.P.C. to bottom, back, and ends of precast concrete cills.

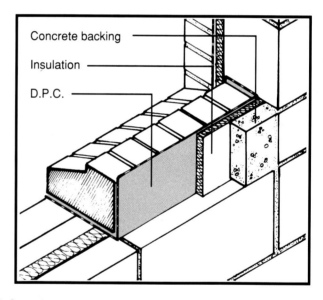

Brick cill:
Provide D.P.C. to bottom, back, and ends of brick cills.

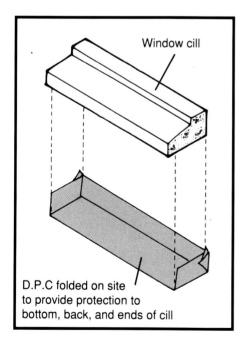

Jamb detail:
Ensure that frames are installed plumb, level, square and fitted to acceptable tolerances. Frame **must** be installed behind the outer leaf.

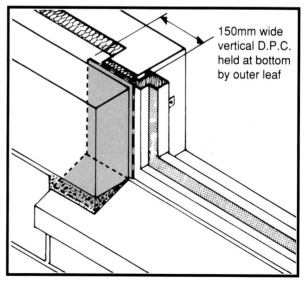

DAMP PROOF COURSES
continued.

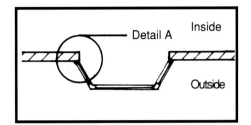

Bay window construction

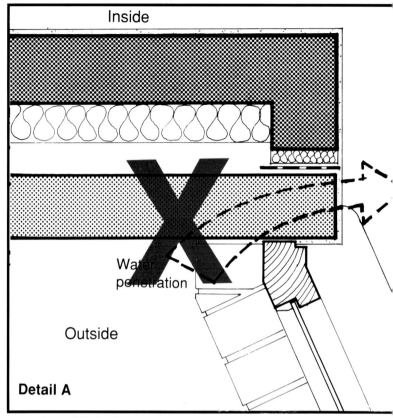

Detail A

Location of D.P.C in bay window construction in the above sketch shows an incorrect location of the D.P.C. as it allows moisture from the outside to penetrate to the inside of the building.

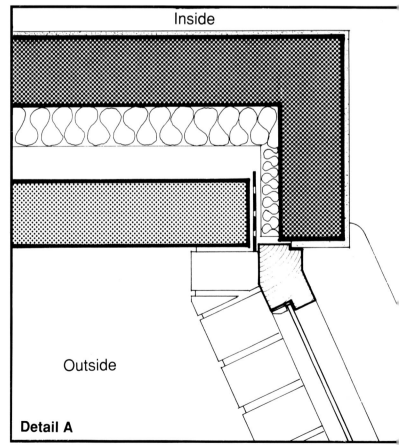

Detail A

Correct location of D.P.C. where dampness cannot penetrate to inside.

DAMP PROOF COURSES
continued.

The results of the absence or incorrect installation of damp proof course and trays.

Pressed metal lintels

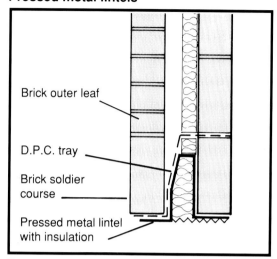

Brick outer leaf

D.P.C. tray

Brick soldier course

Pressed metal lintel with insulation

Provide stepped d.p.c's above all openings in cavity walls.

Prestressed concrete lintels

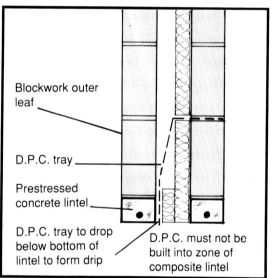

Blockwork outer leaf

D.P.C. tray

Prestressed concrete lintel

D.P.C. tray to drop below bottom of lintel to form drip

D.P.C. must not be built into zone of composite lintel

Provide stepped D.P.C's to all openings in cavity walls or proprietary cavity tray D.P.C's..

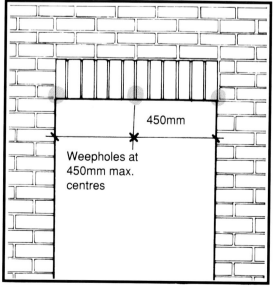

450mm

Weepholes at 450mm max. centres

Where outer leaf is unrendered (brick for example) provide weepholes at 450mm max. centres above opening. Avoid any lodging of mortar droppings in D.P.C. tray.

DAMP PROOF COURSES
continued.

Flat roof and mono-pitch roof abutments of the type illustrated can occur at porches, balconies etc., and require careful detailing of damp-proof courses and flashings.

The cover flashings should be built in under the stepped D.P.C. in all cases.

Flashings should be of non-ferrous metal. Where lead is used flashings should be at least code 4 (colour code blue). Individual lengths of flashing in abutments should not exceed 1800mm with laps not less than 100mm.

Note: Getting D.P.C. installation right is vital because any moisture that falls down the cavity will show as damp on the ceiling of the habitable area of the dwelling.

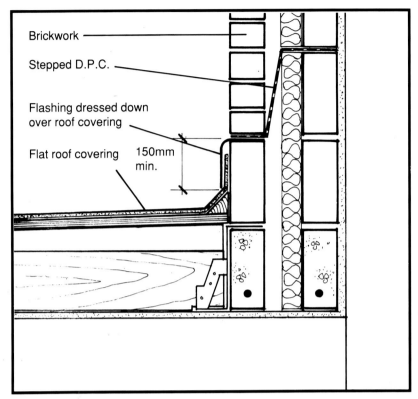

Flat roof abutment.

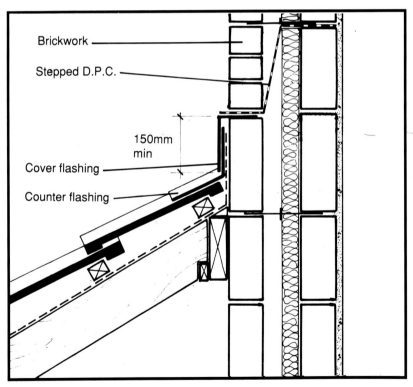

Mono–pitch roof abutment.

CHANGES IN ROOF LEVEL

Changes in roof level between buildings of different heights or arising from steps and/or staggers in building layout can give rise to an external wall at a higher level becoming an internal wall at lower level. It is vital that adequate damp proof-courses are used in such cases. The use of preformed cavity trays in such locations is strongly recommended.

The cavity tray is required as flashings and soakers are not sufficient. Below the roof line the external leaf becomes an internal wall, so it is necessary to prevent any moisture that penetrates the outer leaf from running down the building, as illustrated. This is essential in brickwork outer leaves. For rendered walls see page 69.

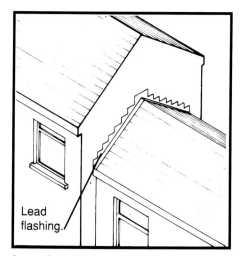

Location of cavity trays.

Lead flashing.

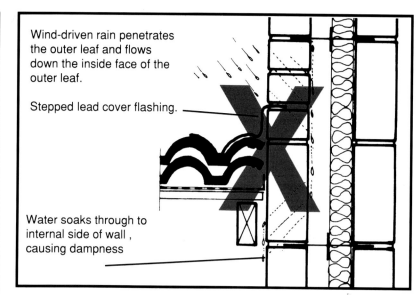

Wind-driven rain penetrates the outer leaf and flows down the inside face of the outer leaf.

Stepped lead cover flashing.

Water soaks through to internal side of wall , causing dampness

No cavity tray; this construction method is incorrect.

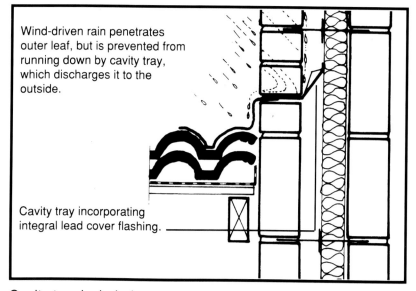

Wind-driven rain penetrates outer leaf, but is prevented from running down by cavity tray, which discharges it to the outside.

Cavity tray incorporating integral lead cover flashing.

Cavity tray included.

Cavity trays with integral lead flashing.

CHANGES IN ROOF LEVEL
continued.

Cavity tray range

Catchment tray: This is the first tray to be built into the outer leaf. It should have upstands at both ends and incorporate some sort of discharge unit integrally designed or by other means for any water that may run down from higher up. The upstands prevent water from entering the cavity.

Intermediate trays: These trays are handed, with an upstand at one end, and it is recommended they have 1 weephole per cavity tray (If this is not possible weepholes at 1m centres max.) to divert water safely to the roof covering. Each tray should overhang the one immediately below by min. 100mm.

Apex tray: Identical to the catchment tray. However, the shape of the attached lead flashing is different.

When using cavity trays the following points should be considered:

◆ Use purpose made self weeping trays.

◆ Only build trays into the outer leaf.

◆ Tray dimensions may vary depending on roof pitch.

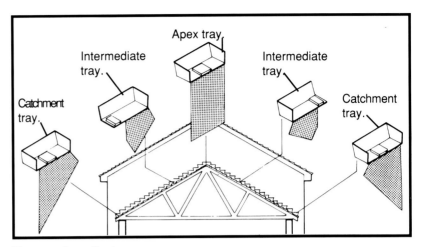

Types of cavity tray.

When built in each unit discharges individually onto the roof slope. This avoids a build up of water flow towards the bottom of the run.

SETTING OUT

Great care should be exercised when setting out and the manufacturer's instructions followed carefully.

Note: Where it is necessary to cut bricks to facilitate equal spacing of the trays the cut brick should be covered by the lead flashing from the tray above.

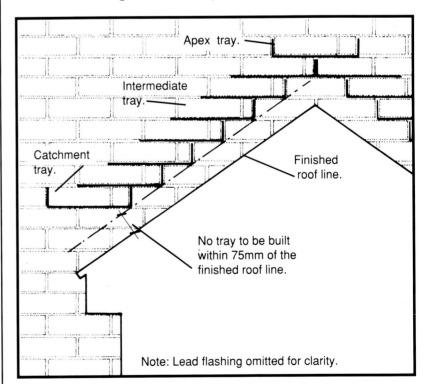

Setting out.

CHANGES IN ROOF LEVEL
continued.

Rendered Walls.
The same trays that are used for brick walls can also be used for rendered blockwork walls. To facilitate this the region in which the trays are to be inserted must be constructed in concrete brickwork, this ensures proper spacing of the trays.

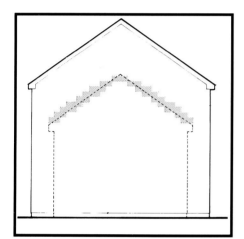

The bricks will be hidden by the flashing and the render, and the render can also partly conceal the lead. Rendering should not be applied directly to the flashings, as this restricts movement and may cause splitting of the flashing or detachment of the rendering. Expanded metal mesh should be fixed to the blockwork, extending down to a bellcast stop bead 75mm min. off the finished roof line. This provides a key for the render and enables the lead to move. Fixings must not pass through the lead. The render will block the weepholes of the intermediate trays so it is vital to ensure that a weephole is provided to the catchment tray to ensure any water collected can drain freely onto the roof surface.

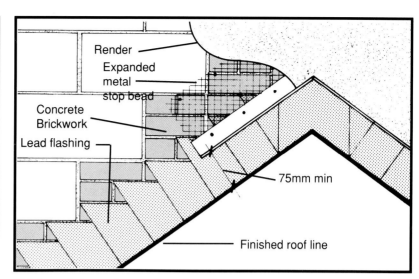

Brickwork trays in rendered walls:
Note the concrete brickwork where the trays are inserted.

Note: Notice the patent galvanised/stainless steel render stop bead – do not nail the render stop bead to the lead flashing.

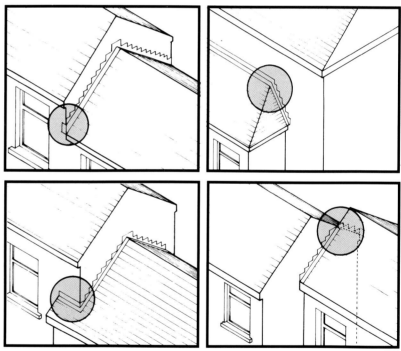

Key junctions: The junctions illustrated above require special care and attention when being constructed. It is recommended that proprietary purpose made products should be used and the product manufacturer consulted to ensure proper installation.

Note:
It is not normal practice to incorporate cavity trays in rendered blockwork walls. However in areas where there may be exposure to high levels of driving rain the use of such trays should be considered.

Catchment tray incorporating lead flashing. The first tray to be built in.

Catchment tray built into outer leaf and mortar bedding provided for the next brick course. Note the weephole.

No tray to be built within 75mm of the finished roofline.

Intermediate tray. This is the next tray to be built in and should overhang the tray below by 100mm min. Note the cut bricks these will be covered by the lead flashing.

Apex tray. Note its location in relation to the ridge.

The completed cavity tray built up the lead flashing is dressed over the roof finish to ensure proper weathering.

CAVITY INSULATION

The fundamental principle of cavity wall construction is that the cavity prevents moisture moving from outside to inside.

NHBGS RULES:

1   WHERE INSULATION MATERIAL IS BUILT INTO THE CAVITY DURING CONSTRUCTION BY THE MASON, AND THE OUTER SKIN IS FINISHED IN FACING BRICK OR FAIR FACED BLOCKWORK, A CLEAR AIRSPACE OF 40MM IS ALWAYS REQUIRED.

2   CAVITY–FILL INSULATION, INSTALLED AFTER CONSTRUCTION OF THE WALL IS NOT PERMITTED WHERE THE OUTER SKIN IS FINISHED IN FACING BRICK OR FAIR–FACED BLOCKWORK.

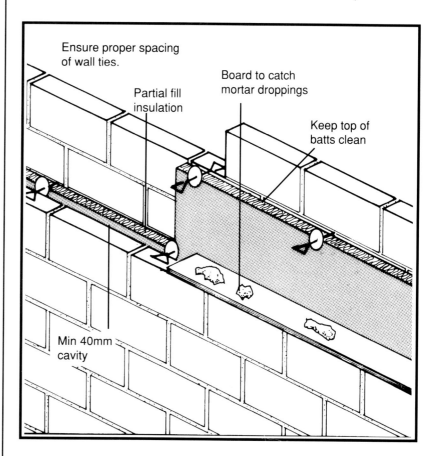

Ensure proper spacing of wall ties.

Partial fill insulation

Board to catch mortar droppings

Keep top of batts clean

Min 40mm cavity

Building in insulation:
Use wall ties to suit the type of insulation being used. Keep top of batts free of mortar droppings. Ensure insulation is held tight against inner leaf. Vertical joints between insulation batts should be tightly butted. See page 223 for guidance on storage of insulants.

INSULATION AROUND OPENINGS

Head of openings:
details to ensure continuity of Insulation.

Note:
D.P.C. or flashings not to be built into the zone of composite action of the lintel.

To prevent 'cold bridging' insulation should be present around the perimeter of openings as well as in the general wall areas.

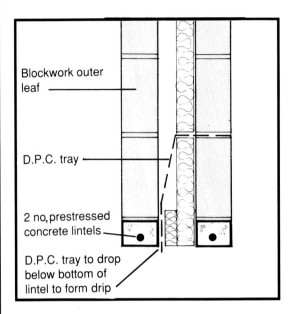

Blockwork outer leaf

D.P.C. tray

2 no. prestressed concrete lintels

D.P.C. tray to drop below bottom of lintel to form drip

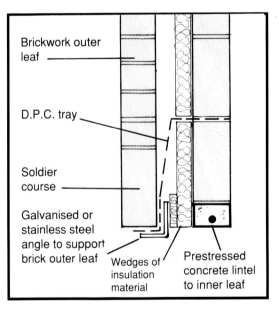

Brickwork outer leaf

D.P.C. tray

Soldier course

Galvanised or stainless steel angle to support brick outer leaf

Wedges of insulation material

Prestressed concrete lintel to inner leaf

Note: Angle sizes will vary with span – See page 56 for guidance.

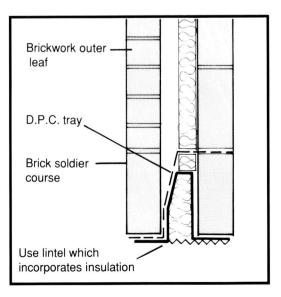

Brickwork outer leaf

D.P.C. tray

Brick soldier course

Use lintel which incorporates insulation

Note: ensure all lintels bear on full solid block where possible.

INSULATION AROUND OPENINGS
continued.

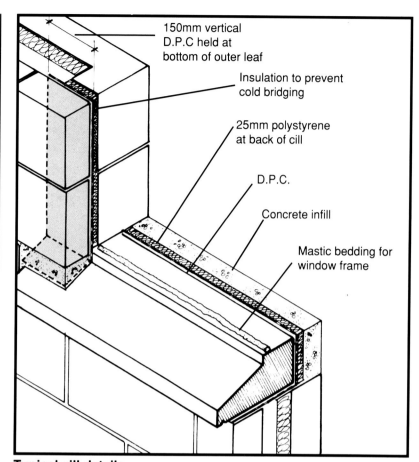

Typical cill details.

Note: Mastic gunned to cill immediately before fixing window frame.

Labels on figure:
- 150mm vertical D.P.C held at bottom of outer leaf
- Insulation to prevent cold bridging
- 25mm polystyrene at back of cill
- D.P.C.
- Concrete infill
- Mastic bedding for window frame

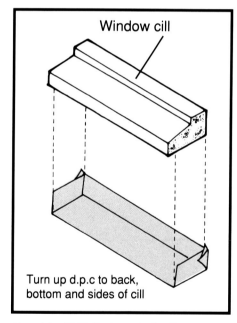

Window cill

Turn up d.p.c to back, bottom and sides of cill

Provide D.P.C. tray to bottom, back and sides of cill.

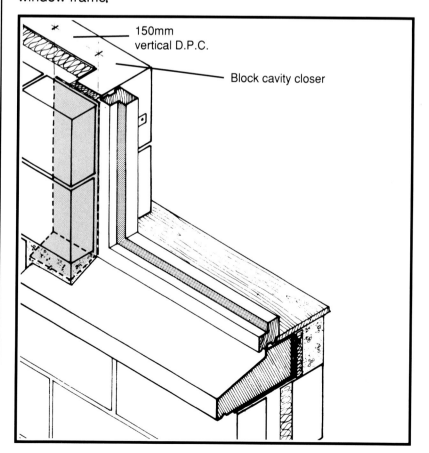

Labels on figure:
- 150mm vertical D.P.C.
- Block cavity closer

Typical jamb treatment.

**BRICKWORK AND
BLOCKWORK GENERALLY**

To avoid the risk of rain
penetration through brickwork,
care is needed in bricklaying,
especially to ensure that all
vertical and horizontal, joints are
properly filled.

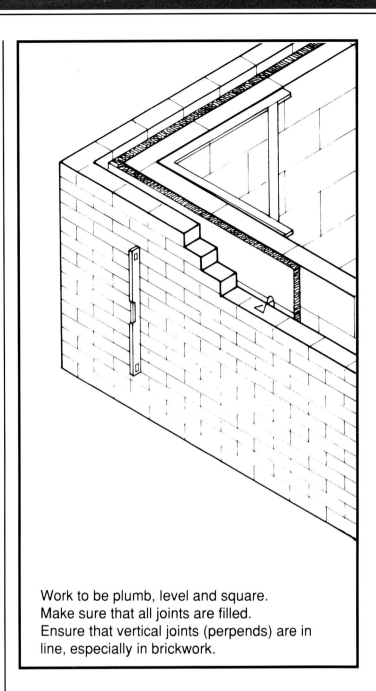

Work to be plumb, level and square.
Make sure that all joints are filled.
Ensure that vertical joints (perpends) are in
line, especially in brickwork.

Keep brickwork free from mortar splashes.

**BRICKWORK AND
BLOCKWORK**
continued.

Protect brickwork against the
weather, at breaks in the work
sequence, (e.g. overnight and at
weekends), and when walls overhead
are being rendered.

All blockwork and brickwork should
be carried up regularly with no part
at any time to be more than 1m
higher than any other. This also
applies to the leaves of cavity
walls.

Blockwork, especially in the lower
areas, carries all the weight of the
house. It is essential that all walls
are properly and securely bonded
to support the load.

**"TOOTHING IN" NEVER
ACHIEVES APPROPRIATE
STRENGTH.**

Carry up all blockwork/brickwork (external walls,
internal walls and chimney stack) together.

BRICKWORK AND BLOCKWORK
continued.

The layout of openings in walls should be carefully considered to avoid problems of over-stressing of masonry. This is a matter for the building designer at the design stage.
Useful guidance in this area is given in Technical Guidance Document A of the Building Regulations.

Particular attention should be given to openings in 'buttressing walls' as defined in Building Regulations Technical Guidance Document A.
An example of the provisions of the Technical Guidance Document is illustrated below.

The information illustrated on this page can also be found in the NHBGS publication "Housing and the Building Regulations".

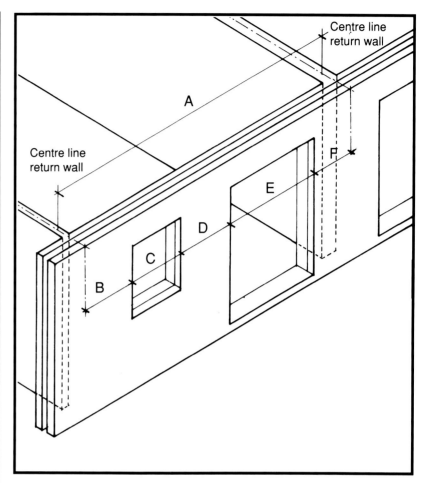

Typical external wall openings.

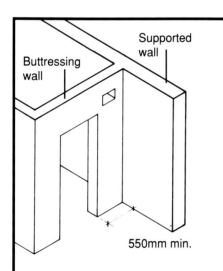

There may be one opening or recess not more than 0.6m² at any position.

Any opening or recess greater than 0.6m² should be at least 550mm from the supported wall.

Openings in buttressing walls

Examples of recommendations given in Technical Document A of the Building Regulations 1991 for 2-storey house.

1   The combined width of the openings should not extend beyond two thirds the length of the wall.

i.e. ope C + ope E should not be greater than $\dfrac{2A}{3}$

2   No individual opening should exceed 3m in width.

i.e. width C or width E not greater than 3m

3   A pier between an opening and a return wall should be equal to at least $\dfrac{\text{ope width}}{6}$

i.e. Pier B should be equal to or greater than $\dfrac{C}{6}$
(pier B min. 550mm if wall is a buttressing wall.)

i.e. Pier F should be equal to or greater than $\dfrac{E}{6}$
(pier F min. 550mm if wall is a buttressing wall.)

4   A pier between two openings should be greater than or equal to ¹/₃ the sum of these two openings

i.e. D should be greater than or equal to $\dfrac{C+E}{3}$

PIERS AND OPENINGS

Narrow piers between openings give rise to high stresses in masonry and the width must be controlled.

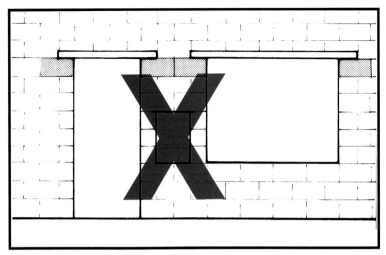

Consider carefully the locations of elements such as meter boxes to avoid problems of over stressing. Do not locate a meter box in the location shown above, the gable wall is a good location for the meter box. The head of the meter box is generally supported by galvanised or stainless steel angle or concrete lintel. A stepped D.P.C. must be provided at the head also. See page 72 for stepped D.P.C. arrangements at heads.

The forming of chases for services must be controlled to avoid excessive weakening of the wall.
Chases must not be back to back.

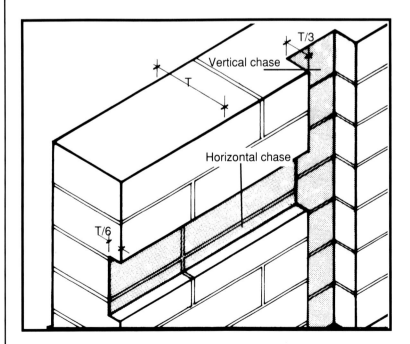

Maximum depth of vertical chase to be one-third of wall thickness.
Maximum depth of horizontal chase to be one-sixth of wall thickness.

INTRODUCTION

The properties required of external rendering are that it must be durable, resist moisture penetration, weather uniformly and in addition possess an attractive appearance.

Durability:
Durability depends on a number of factors, including proper detailing, the degree of exposure of the building, the proportions of the mix, the bond between the rendering and the wall, and the standard of workmanship.

Resistance to moisture:
Resistance to moisture penetration depends primarily on the detailed design of the building and the mix used for rendering. Water is less likely to penetrate through rendering of an absorbent character than through cracks in the rendering caused by the use of a strong dense mix.

Uniform weathering:
Uniform weathering depends mainly upon the texture of the finished surface, the nature of the mix, and structural details such as cills, copings, eaves etc..

Specification for rendering:
Blockwork should be scudded with a mix of $1:1\frac{1}{2}$ to 2, cement : sharp sand, and finished with two coats of render mixed in the proportions 1:1:6, cement : lime : sand or 1:5 or 6 cement : sand with plasticiser added. Plasticiser must be used in strict accordance with the manufacturer's recommendations. Care should be taken to avoid overdosing or overmixing which may result in excessive air content leading to loss of strength and durability.

The undercoat should be between 8mm and 16mm thick. The final coat should be thinner than its undercoat, generally less than 8mm.

Volume batching:
Volume batching should be carried out using properly constructed gauge boxes.

Scud coat:
Rendering should not be applied to blockwork without the provision of an adequate mechanical key to hold the rendering both during and after its application. Shrinkage of the rendering sets up stress between itself and the background.

The bond must be strong enough to resist separation. The use of a scud coat is recommended as a method of providing a key.

The scud coat should be mixed with just sufficient water to give the consistency of a thick slurry. It is best applied by throwing with a hand scoop to a thickness of approx. 3–5mm. The surface should be dampened periodically until hardened and then allowed to dry.

Undercoat:
The undercoat should be combed or scratched after it has been left long enough to set firm, care being taken to leave the scratched marks sufficiently deep to provide a key for the following coat, but not so deep as to penetrate through the undercoat.

Final coat:
Before applying the final coat the undercoat should be allowed to harden and dry out sufficiently to provide adequate suction. Where necessary the suction of the undercoat should be reduced by uniformly wetting. In plain finishes the final coat should be thinner than its undercoat (generally less than 8mm) and should be applied with a suitable float e.g. wood. A steel trowel should not be used, and over working should be avoided.

Successive render coats should be specified as being no stronger than the previous coat or background.

Float ready plaster:
Float ready plaster is a premixed ready to use mortar, specially formulated for external rendering and internal plastering. It is produced in mixing plants and delivered to site in quantities to meet customers' requirements. Its use must be in strict accordance with the manufacturer's instructions.

TYPES OF FINISH

◆ **General**

In general, textured finishes are less liable to crack and craze than a plain finish; also, any cracks that do develop are less likely to be obtrusive. Textured finishes are easier to bring to a uniform appearance; this is an important consideration when the rendering is coloured. Although offering more lodgement for dirt, a rough texture tends to even out any discoloration which makes the dirt less apparent than with smoother finishes. Also, the distribution of the flow of rainwater over a textured surface reduces the risk of penetration through the rendering.

The finishes for external renderings fall into two broad categories: trowel finishes and thrown finishes.

◆ **Trowel finishes**

(i) Plain finishes

Plain finishes should be achieved by using a wooden float. This type of finish requires a high standard of workmanship to minimize the risk of crazing and irregular discoloration. A steel trowel should not be used for finishing an external render.

(ii) Scraped or textured finishes

Textured finishes should be achieved by working the surface of the freshly – applied final coat with a trowel or other hand tool or, alternatively, a textured finish can be applied direct from the nozzle of a rendering machine. Scraped finishes should be obtained by allowing the final coat to harden for several hours and then by scraping it with a suitable tool.

◆ **Thrown finishes**

(i) General

Under severe conditions of exposure, thrown finishes are generally more satisfactory than trowel applied finishes in respect of weather resistance, durability and resistance to cracking and crazing. Roughcast and drydash finishes are normally used on strong backgrounds. Mix properties will need to be modified for use on weaker backgrounds.

(ii) Roughcast

A roughcast finish should be achieved by throwing on the final coat of rendering as a wet mix and leaving it un–trowelled. The "roughness" is determined by the shape and size of the coarse aggregate in the mix.

(iii) Drydash

A drydash finish should be produced by throwing crushed rock chippings or pebbles on to a freshly – applied mortar layer (buttercoat) using a small shovel or scoop and leaving it exposed.

◆ **Curing**

It is essential that a newly rendered surface be prevented from drying out too rapidly, although protection from the sun and wind, or spraying with water may only be necessary in hot and dry weather. Spatterdash, however, should be wetted down an hour or so after application to ensure adequate hydration.

Each coat should be allowed to shrink and dry out as long as possible normally for a period of several days, before the subsequent coat is applied. In cold wet weather this process will take considerably longer than in warm dry weather.

Consideration should be given to more stringent than normal protection of coloured rendering and decorative finishes in hot weather, wind, rain, or other adverse conditions.

◆ **Resistance to crazing**

Crazing results from differential shrinkage of the surface of the rendering in relation to its interior. The cracks formed are narrow and generally do not extend far below the surface: they may, however develop into shrinkage cracks. Cement–rich steel–trowelled finishes are particularly liable to craze; on the other hand, leaner mixes with a scraped, textured, or other rough finish are resistant to this defect.

TYPES OF FINISH
continued.

The risk of crazing should be minimized by:

◆ The use of properly graded sand, in particular the avoidance of an excessive proportion of very fine material e.g. silty dirty sand.

◆ The use of a mix which is relatively lean in cement.

◆ The avoidance of over working, which causes an excessive laitance to be drawn to the surface.

◆ The avoidance of too rapid drying out of the final coat.

INTRODUCTION

Most buildings develop cracks in their fabric, usually soon after construction, sometimes later. Much of the cracking is superficial, easily repaired and unlikely to recur to any great extent. Only rarely does cracking indicate serious structural failure. Whether superficial or not, much can be done to minimise cracking by recognising that movement of building materials and components is inevitable and should be allowed for in design.

The following pages address the issue of cracking due to shrinkage and / or expansion, and give guidance on provisions to minimise its occurrence.

There are many reasons for cracking which include settlement of foundations, proximity of trees and spreading of roof structures.

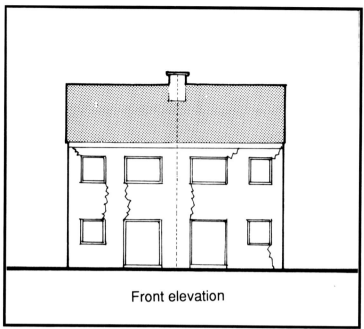

Front elevation

Typical cracking patterns

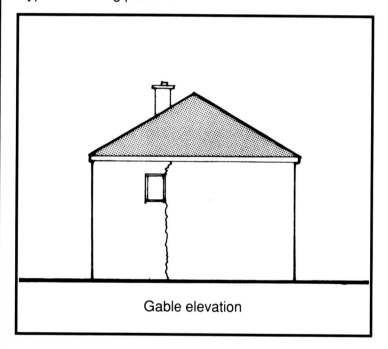

Gable elevation

Cracking patterns commonly encountered in houses built of concrete blocks.

Concrete blocks which are manufactured in the open air have a high moisture content. If not cured properly, they shrink when built into a house, and cracking at the weakest areas such as over and under window openings results. Blocks should not be used until they have been cured adequately. Check that blocks have been stored for at least 4 weeks prior to delivery to minimise the risk of cracking.

Use properly – cured blocks to reduce the risk of cracking.

MOVEMENT JOINTS

Long lengths of wall are prone to contraction and expansion with cooling and heating as well as drying. If account is not taken of this there is a risk of cracking. To reduce this risk, the incorporation of movement joints is necessary. In the case of concrete blockwork these will be contraction joints and in the case of clay brickwork expansion joints.

Codes of practice recommend that spacings of movement joints should be referred to by the building designer. It is not standard practice to incorporate movement joints in semi–detached houses. Based on traditional construction the minimum recommendation of HomeBond is that in terraces of three or more houses, joints should be built in every two houses unless specified more frequently by the designer.

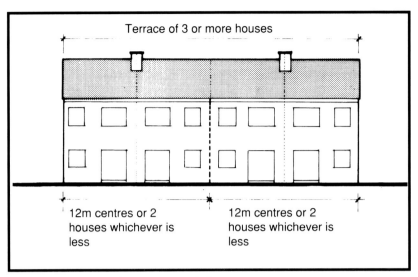

Typical location of movement joints.

Note: Movement joints are provided to front and back of terraces. Joints are often concealed behind rainwater pipes.

MOVEMENT JOINTS IN CLAY BRICKWORK

Provide ties at 225mm vertical centres on each side of the joint. The joint should be 18mm wide minimum.

The joint should be pointed with either:

one part polysulphide sealant or one part low modulus silicone, on a backing of suitable material which may include flexible closed cell polyethylene, in cord or sheet form.

The joint illustrated on this page (in the outer leaf only) would apply to relatively short terraces with brick outer leaves i.e. terraces of up to four houses. For longer terraces, see page 84.

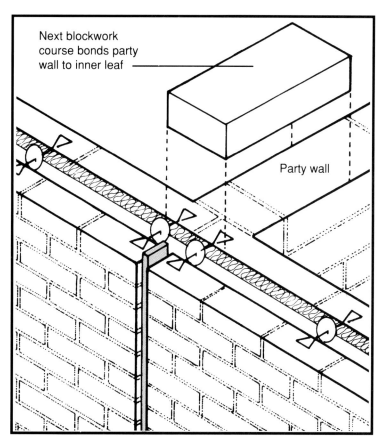

Typical movement joint in clay brickwork outer leaf.

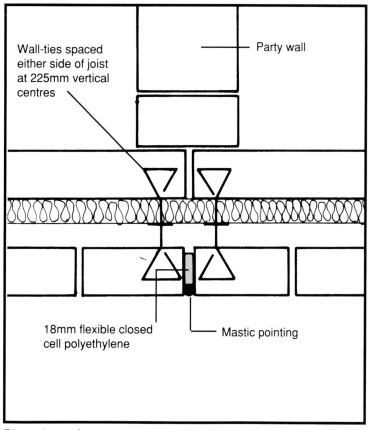

Plan view of movement joint in clay brickwork outer leaf.

MOVEMENT JOINTS IN CLAY BRICKWORK
continued.

In terraces comprising five or more houses, the provision of control joints to both inner and outer leaves should be considered. This sketch illustrates such a detail, and the sketches on the following pages give further details of such a joint.

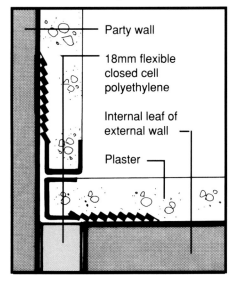

Plan detail of internal corner finished with preformed galvanised or stainless steel plaster stops.

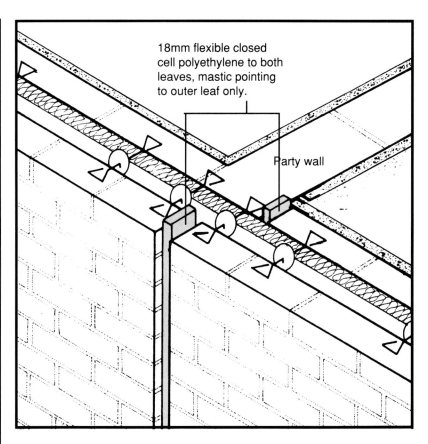

Movement in clay brick outer leaf and blockwork inner leaf. This applies to a terrace of 5 or more houses.

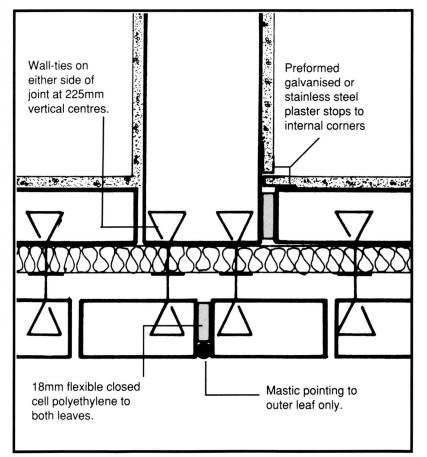

Plan view of movement joint in clay brick outer leaf and blockwork inner leaf.

MOVEMENT JOINTS IN CLAY BRICKWORK
continued.

As an alternative to the provision of a joint in the inner leaf of the type shown on the previous page, another approach is to construct the party wall at the movement joint as a cavity wall, with 50mm cavity.

Note:
Where party wall is of cavity construction use double triangle wall ties across cavity of party wall.

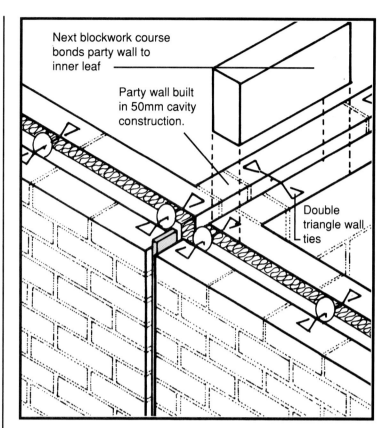

Movement joint to clay brickwork outer leaf.

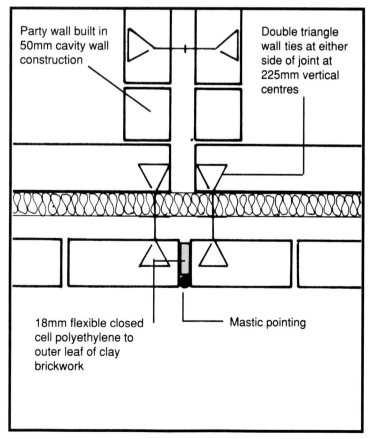

Plan view of movement joint to clay brickwork outer leaf.

MOVEMENT JOINTS IN CLAY BRICKWORK
continued.

In the case of stepped blocks / terraces the movement joints are best accommodated in the internal corner of returns as illustrated here.

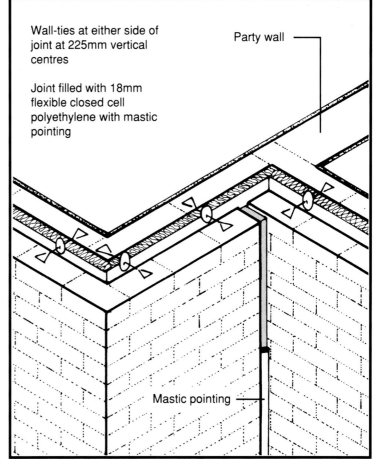

Wall-ties at either side of joint at 225mm vertical centres

Joint filled with 18mm flexible closed cell polyethylene with mastic pointing

Party wall

Mastic pointing

Movement joint in a stepped terrace.

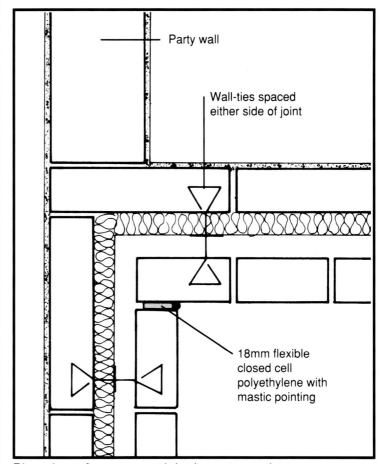

Party wall

Wall-ties spaced either side of joint

18mm flexible closed cell polyethylene with mastic pointing

Plan view of movement joint in a stepped terrace.

MOVEMENT JOINTS IN CONCRETE BLOCKWORK

As with brickwork, block outer leaves are prone to movement and, in particular, shrinkage. Movement joints are therefore recommended in any block or terrace of three or more houses where joints in outer leaves should be at 12m maximum centres, unless specified more frequently by the designer.

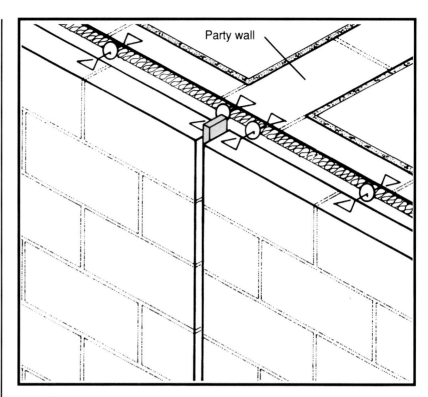

Detail of movement joint in blockwork outer leaf.

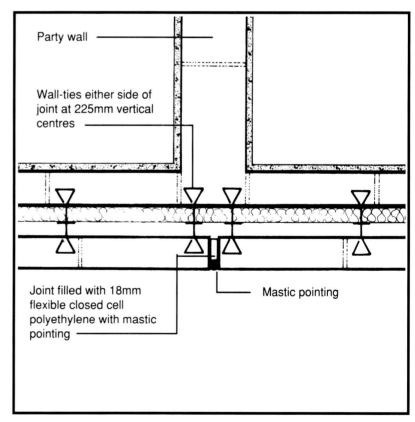

Plan view of movement joint in blockwork outer leaf.

MOVEMENT JOINTS IN CONCRETE BLOCKWORK
continued.

Movement joints are not normally necessary in the inner leaf of a cavity wall, but consideration should be given at the design stage to providing movement joints to the inner leaf of a cavity wall in terraces of five or more houses.

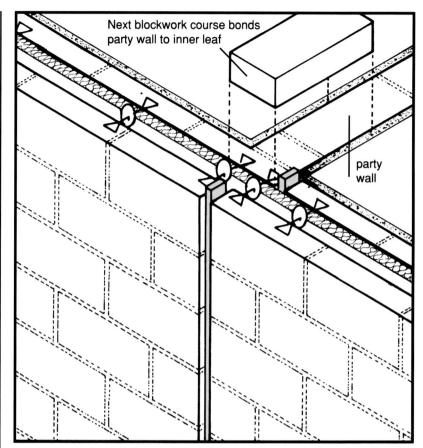

Next blockwork course bonds party wall to inner leaf

party wall

Detail of movement joint in both leaves of concrete block wall.

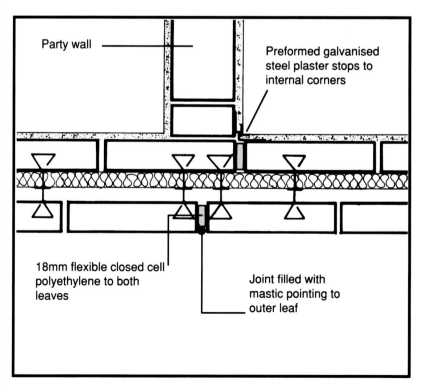

Party wall

Preformed galvanised steel plaster stops to internal corners

18mm flexible closed cell polyethylene to both leaves

Joint filled with mastic pointing to outer leaf

Plan view of movement joint in both leaves of concrete block wall.

MOVEMENT JOINTS IN CONCRETE BLOCKWORK
continued.

Hollow block walls.
As in cavity wall construction where a terrace contains three or more houses, movement joints should be provided at maximum 12m centres unless specified more frequently by the designer.

The sketch opposite suggests a method of incorporating a movement joint into hollow block wall construction.

Note:
Where party wall is of cavity construction use double triangle wall ties across cavity of party wall.

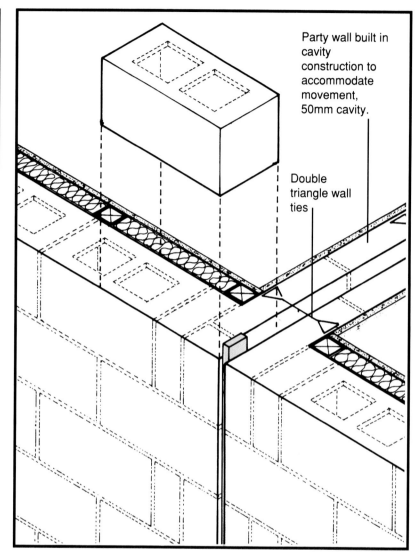

Party wall built in cavity construction to accommodate movement, 50mm cavity.

Double triangle wall ties

Hollow block external wall – movement joint.

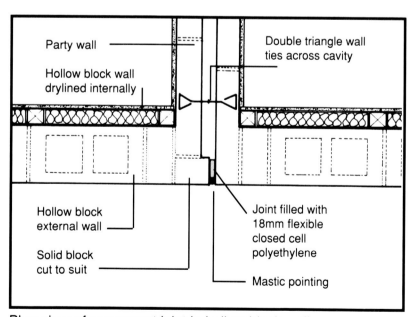

Party wall

Hollow block wall drylined internally

Double triangle wall ties across cavity

Hollow block external wall

Solid block cut to suit

Joint filled with 18mm flexible closed cell polyethylene

Mastic pointing

Plan view of movement joint in hollow block wall.

MOVEMENT JOINTS IN CONCRETE BLOCKWORK continued.

Hollow block walls.
As an alternative to the movement joint shown on the previous page for hollow block walls, the type of joint illustrated on these pages can also be used in terraces containing three or more houses.

The sketches on this and the following page illustrate the sequence of construction of such a joint, and emphasise that wall-ties be provided at every horizontal bed joint to securely tie the party wall to the external wall.

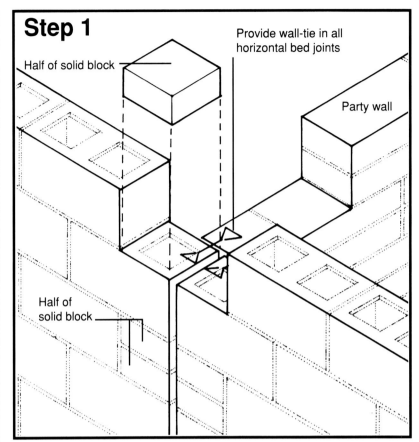

Sequence of construction – hollow block external wall movement joint.

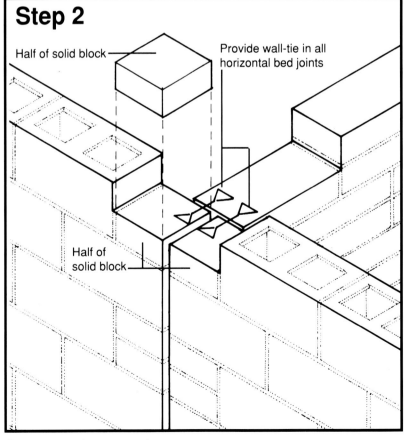

Sequence of construction.

**MOVEMENT JOINTS IN
CONCRETE BLOCKWORK**
continued.

Hollow block walls.

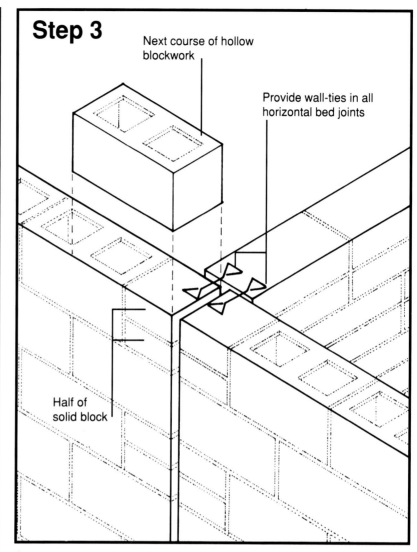

Sequence of construction.

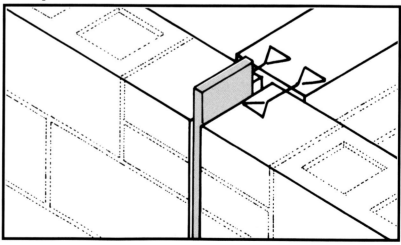

Completed joint filled with 18mm flexible closed cell
polyethylene with mastic pointing

PREFORMED MOVEMENT JOINT BEADS

For use with externally rendered or plastered walls.

The bead consists of two lengths of galvanised or stainless steel plaster stop bead jointed by a PVC extrusion which allows movement across the joint.

The beads are fixed by dabs or by galvanised nails driven through the holes provided to the external surface of the wall across the joint. The face of the PVC extrusion is fitted with a protective tape which is removed after plastering. Ensure that the movement joint bead used is suitable for external use.

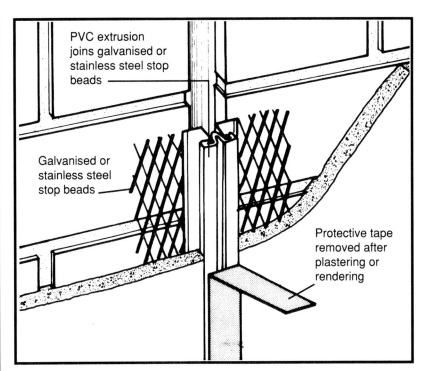

Movement joint bead

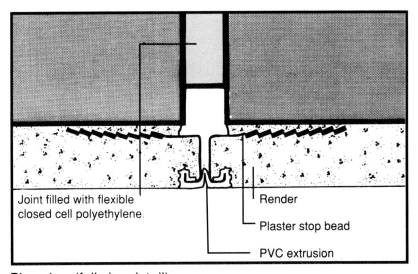

Plan view (full size detail)

BED JOINT REINFORCEMENT

Cracking may be controlled by the use of patent reinforcement in horizontal joints in areas of high stress such as above and below openings. The reinforcement should be long enough, at least 600mm either side of ope, to distribute the high stress to areas of low stress. The reinforcement used in an external wall should be corrosion resistant such as stainless steel and should have at least 20mm cover from the outer wall face.

Wall panel provisions.

It should be noted that the use of movement joints, as recommended in the preceding pages, will not necessarily eliminate completely the occurrence of shrinkage cracking that extends diagonally from the corner of openings in blockwork walls. There is an increased risk of cracking if the length of a blockwork panel exceeds twice the height, and the shape is often more critical than size.

For example a large square gable may remain free of cracks while a low panel below a long window may crack.

In any bungalow built of blocks the incorporation of movement joints is recommended wherever the length of uninterrupted wall exceeds twice the height.

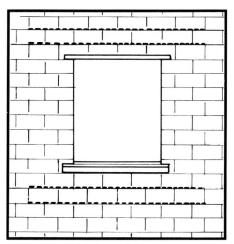

Typical bed joint reinforcement at openings (shown dotted).

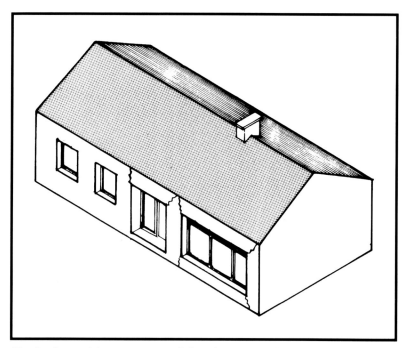

Typical cracking pattern arising from panel proportions.

CORBELLING OF BRICKWORK AND BLOCKWORK

When it is proposed to provide a corbel in wall construction it is important that the extent of the overhang should not exceed 1/3 of the original wall thickness – or in the case of cavity wall construction 1/3 of the thickness of the leaf containing the corbel, as illustrated, unless the corbel detail is designed to take account of the specific circumstances. Design guidance on this matter is given in clause 27.8, BS 5628: Pt. 3: 1985 – Code of practice for use of Masonry.

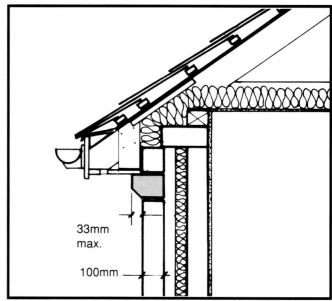

Corbel at eaves level — maximum projection should be 33mm.

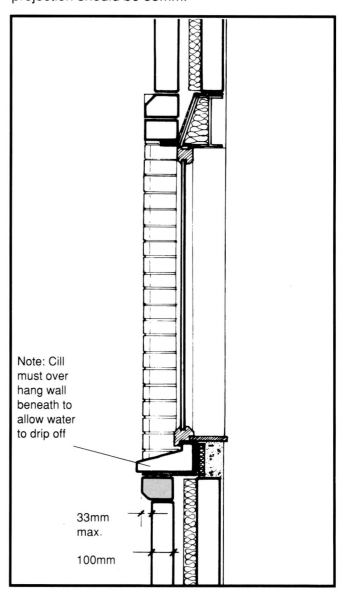

Note: Cill must over hang wall beneath to allow water to drip off

Corbel above and below window ope — maximum projection should be 33mm.

PARTY WALL CONSTRUCTION

Attention to detail is important in this area to ensure avoidance of

1 Excessive sound transmission
2 The risk of fire spread.

The guidance in the Building Regulations Technical Guidance Document E for sound insulation states that joists at right angles to a party wall should be fixed with joist hangers.

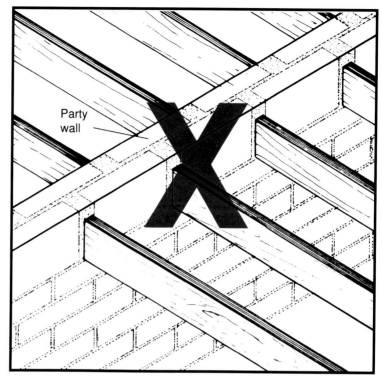

DO NOT BUILD JOISTS INTO PARTY WALL.

Where possible joists should span from front to back of house. Where this is not possible, and joists must span on to party wall, use joist hangers. All joints to be filled to avoid excessive noise transmission and risk of fire spread.

Note:

1 Ensure hangers are correct size.

2 Ensure hangers are tight against blockwork.

3 Ensure joists are notched into hangers.

4 Ensure joists are securely nailed into hangers.

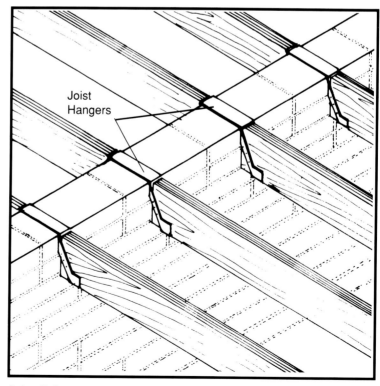

Joist fixing at party wall.

PARTY WALL CONSTRUCTION
continued.

Forming top of party wall to slope
of roof.

Ensure that party wall is
completed along the line of the
slope of the roof – this is to
reduce the risk of fire spread.

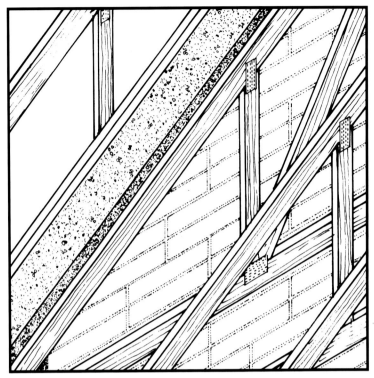

Cut blocks and pack with mortar raked off smooth to
follow line of roof. Leave 25mm approx. gap between
top of mortar and top of rafter to allow for roof
settlement. This gap will be packed with compressible
material (e.g. mineral wool) at the roofing stage. See
also pages 140 and 141 for guidance on fire stopping
the space between sarking felt and roof covering.

Note: Ensure that there are no holes or opes
in party wall.

Carry fire stopping to
underside of roof
covering.

Mortar bedding
between battens or
other suitable
material e.g. mineral
wool.

Mineral wool

Fire stop along top of
wall to underside of
felt with suitable non-
combustible material.

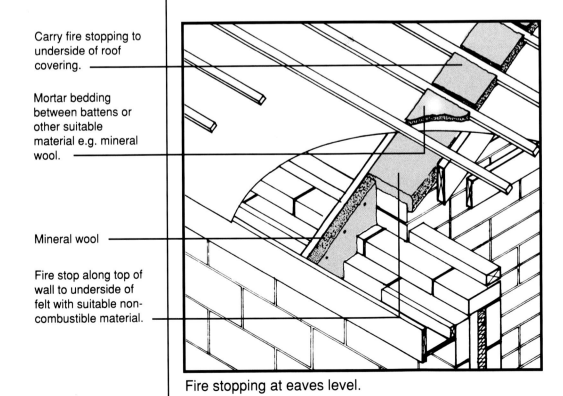

Fire stopping at eaves level.

INTRODUCTION

Mortar quality is an important consideration in the design and construction of brickwork and blockwork. A good mortar is easy to work with, not likely to cause efflorescence in brickwork and provides the required strength necessary to support the brickwork or blockwork. The type of mortar used may also have an effect on shrinkage / expansion cracking. Weak mortars will allow a very slight change in dimensions in each joint where strong mortar will hold the units together and possibly eventually result in a large crack in one location.

Mortar is made from an aggregate, binder and water. The aggregate is sand with suitable grain consistency and the binder is usually a mixture of lime and cement. Other ingredients can be added such as plasticiser in place of lime or colouring agents to achieve the required colour in the case of brickwork. Plasticising agents are often used instead of lime to give improved workability. These additives entrain small air bubbles which act as a lubricant. The air bubbles serve to increase the volume of the binder paste, filling the voids in the sand and this improves the working qualities. Several types of plasticiser are available and should only be used in accordance with manufacturer's instructions.

A good mortar should cling to the trowel, spread easily and not stiffen too quickly. The basic rule for mortars is: cement provides strength, lime or plasticisers improve the workability.

Generally about one volume of binder is needed for three volumes of sand to give a workable mix but cement mortar of this kind is stronger than necessary for most uses. For weaker mortars, lime or plasticisers are needed to maintain workability.

TYPES OF MORTAR

Cement : sand mortar
Adequate strength in the fully hardened mortar, combined with a rapid development of strength in the early stages, is most conveniently attained by the use of Portland cement, but it is not practicable to adjust the strength simply by varying the ratio of cement to sand, because lean mixes of cement and sand are harsh and unworkable.

Cement : lime : sand mortar
Mortars made with appropriate proportions of Portland cement (including sulphate–resisting Portland cement) and lime, take advantage of the useful properties of each. Cement : lime : sand mortars are designed on the principle that part of the cement is replaced by an equal volume of lime so that the binder–paste still fills the voids in the sand. In this way good working qualities, water retention, adhesion and early strength can be secured without the mature strength being too high. The lime used should be non–hydraulic (high calcium or magnesian) or semi hydraulic.

Air–entrained (plasticised) mortar
Mortar plasticisers which entrain air in the mix provide an alternative to lime for imparting good working qualities to lean cement : sand mixes. In effect, the air bubbles serve to increase the volume of the binder paste, filling the voids in the sand, and this correspondingly improves the working qualities.

Ready–to–use retarded cement : lime: sand mortar and cement : sand mortar
When ready–to–use retarded cement : lime : sand mortar and cement : sand mortar are used, care should be taken to follow the manufacturer's instructions for their use.

Mortar mix
The designer should select the mortar designation with reference to the structural requirements and the degree of exposure, taking into account the type of masonry unit, the type of construction and the position in the building and the possibility of early exposure to frost.

The mortar mixes indicated on the following page are suggested to provide the most suitable mortar that will be readily workable to allow the block/brick layer to produce satisfactory work at an economic rate, to be sufficiently durable and to be able to assist in accommodating strains arising from minor movements within the wall.

Further information is available in:

I.S. 325 Part 1: 1986: Structural use of Unreinforced Masonry

I.S. 406: 1987: Masonry Mortars.

Mortar mixes using ordinary Portland or
sulphate resisting cements when required.

Location	Recommended Cement : lime : sand mix	Recommended Cement : sand mix with plasticiser
General wall area above D.P.C. Brickwork and fairfaced blockwork only in areas of very severe exposure.	1 :½ : 4 to 4½	1 : 3 or 4
General wall area above D.P.C. Brickwork and blockwork: other exposure categories.	1 : 1 : 5 or 6	1 : 5 or 6
Below D.P.C. level and chimney stacks, cappings, copings, cills and free standing walls.	1 :½ : 4 to 4½*	1 : 3 or 4

* Also recommended in areas of high water
run–off in brickwork and fairfaced blockwork,
e.g., under large expanses of glazed area.

INTRODUCTION

Fireplaces and chimneys are a standard item in house construction. Fireplaces can be constructed to accommodate an open fire or closed appliance, of which there are a variety in common use. For a fireplace and chimney to function correctly a number of basic items of good building practice need to be executed. The following pages set out these basics – careful adoption of these practices will avoid the risk of costly remedial measures.

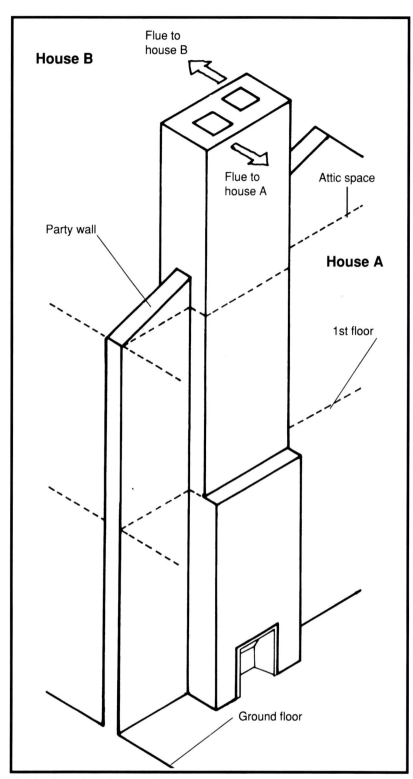

Typical back to back chimney arrangement from ground floor through to roof level.

FIREPLACE RECESSES

Party wall
Fireplaces on party wall.
When fireplaces are on a party wall the thickness of the wall between the fireplaces must be at least 200mm.

Note:
Where two fireplaces occur within the same house the thickness of the dividing wall between the two fireplaces must be at least 100mm.

Where pipes pass from boiler through opes in blockwork, the opes should be fully sealed to prevent smoke escape into the pipe duct as this is frequently a cause of smoke ingress to first floor rooms.

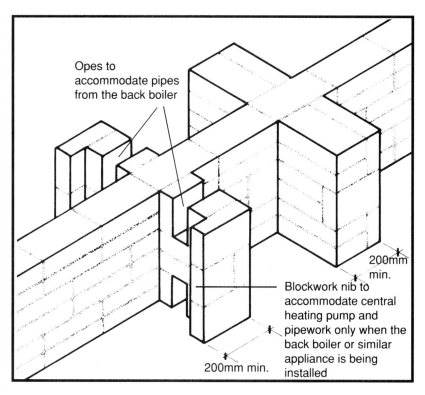

Fireplace recess in party wall (back to back)
Note: Blockwork to chimney stacks must be carried up with all other blockwork.
Subsequent "toothing in" is not permitted.

Fireplaces on external walls.

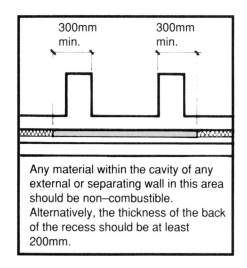

Any material within the cavity of any external or separating wall in this area should be non–combustible.
Alternatively, the thickness of the back of the recess should be at least 200mm.

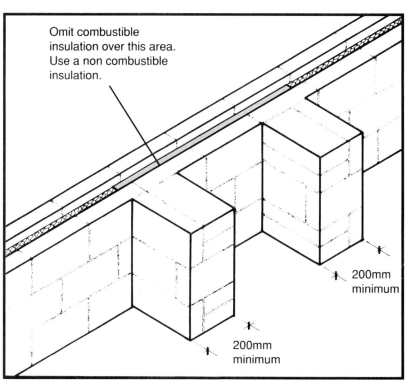

Fireplace recess on external wall .
Maintain cavity where fire place backs on to external wall.

DIMENSIONS OF FIREPLACE OPENINGS

Note: Where a fireplace opening is too high the chimney will not draw properly.
A badly formed throating will cause draught problems.

Flue liners.

Flue liner installation:
Good practice.

1 Use precast flue gatherers.
2 Use spigot and socket flue liners with socket upwards, or rebated flue liner with socket upwards. Liners should be jointed with mortar all round.
3 Pack flue liners as each liner is built in, with:

Cement	Lime	Sand mix.
1	1	12

4 Mix to be wetted with water. Do not use dry fill.
Flue liners must not touch the surrounding blockwork.

DO NOT PACK AROUND FLUE LINERS WITH DRY SAND. USE WETTED CEMENT : LIME : SAND MIX.

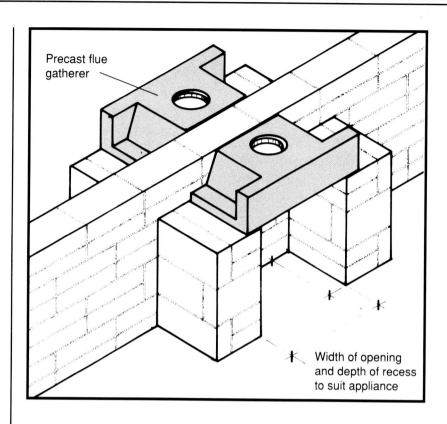

Precast flue gatherer

Width of opening and depth of recess to suit appliance

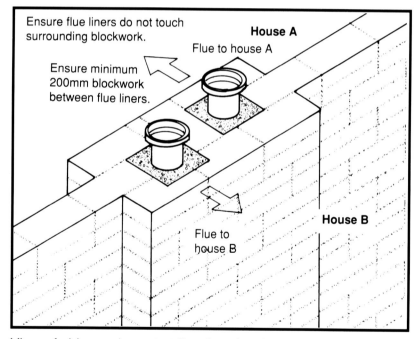

Ensure flue liners do not touch surrounding blockwork.

Ensure minimum 200mm blockwork between flue liners.

House A
Flue to house A

Flue to house B

House B

View of chimney breast at first floor level. Install flue liners as work progresses. Pack around flue liners as each is built in.

The result of packing the flue liner with unsuitable material.

FIREPLACE CONSTRUCTION

For a chimney to draw properly the formation from the top of the fireplace into the flue (the throat) should be properly formed. It is strongly recommended that a pre-formed flue-gathering lintel be used for this purpose to ensure a satisfactory throat arrangement.

Fireplace dimensions should be controlled to avoid the risk of smoking in the finished construction. The height of the fireplace opening measured vertically from the top of the grate to the underside of the flue gathering lintel should not be greater than 550mm.

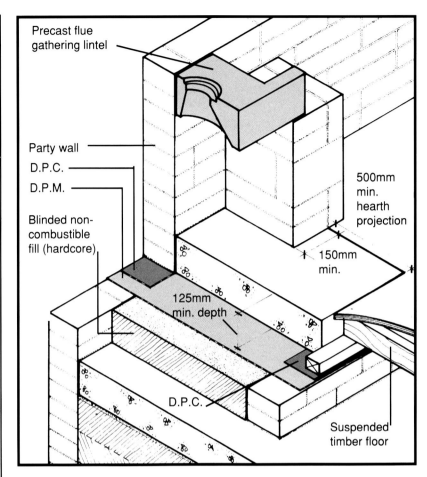

Precast flue gathering lintel

Party wall

D.P.C.

D.P.M.

Blinded non-combustible fill (hardcore)

500mm min. hearth projection

150mm min.

125mm min. depth

D.P.C.

Suspended timber floor

Cut–away view of a typical fireplace and suspended timber floor.

The result of packing the flue liner with unsuitable material.

INSTALLATION OF COOKERS

If an oil fired or solid fuel cooker is installed the installation should be in accordance with the manufacturer's instructions. In particular the connection to the chimney and the chimney construction must be as required by the manufacturer, so that the cooker has sufficient draw and the risk of condensation in the flue is reduced.

Oil fired cookers

In the case of an oil fired cooker, particular attention must be given to the manufacturer's instructions with respect to connection to a standard 200mm flue, as it may be required to line the existing flue with a 150mm rigid or flexible stainless steel flue liner in order to reduce condensation in the flue. Check with the manufacturer of the cooker.

Solid fuel cookers

In the case of solid fuel cookers the correct construction and location of the soot box is essential. Particular attention must be paid to the type of fuel used in order to avoid condensation in the flue: this could result in a liquid tarry substance running down the chimney. If wood or turf with a high moisture content is used then a great deal of moisture is being emitted into the flue and the risk of condensation is greatly increased.

It should be noted that if the appliance is run for extended periods at a low heat, especially when burning wood or peat the flue can cool down to such an extent that vapour in the flue gases may condense. This will make the inside of the flue damp so that the soot sticks to the flue and the tarry mixture formed may drip down into the appliance. It is always best to run at a high rate of heat whenever possible. In the summer period it is better to run the appliance at a high rate to heat the water and then let the fire die and re-light rather than running a low fire continuously.

Ventilation Note

The ventilation requirements for rooms are specified in Technical Guidance Document F of the Building Regulations and in Appendix C of this publication, and must be adhered to in the construction of dwellings. HomeBond also advise that if new windows, whether single or double glazed, are installed in existing houses it is vital that there is sufficient ventilation for the room, particularly those with fireplaces, to ensure that the fire, whether an open or closed appliance, has sufficient air to draw and also that there will not be a build up of the products of combustion which could be a health hazard.

TRIMMING AT FIRST FLOOR LEVEL

Floor joists running parallel or perpendicular to a chimney stack must be trimmed around the stack. Combustible materials **(excluding floorboards, skirting, dado or picture rail, mantleshelf or architrave)** should be separated from a brick or blockwork chimney by at least:

(a) 200mm from a flue, or
(b) 40mm from the outer surface of a brick or blockwork chimney or fireplace recess if blockwork or brickwork is 100mm. Metal fixings in contact with combustible materials should be at least 50mm from a flue.

If using a pair of joists to form trimming / trimmer joists, they should be nailed at 450mm centres 20mm from top and bottom edges of the joist. Alternatively they can be bolted on the centreline at approx. 1m centres. If using this method it is difficult to obtain joist hangers to accommodate the uncommon width of two joists together.

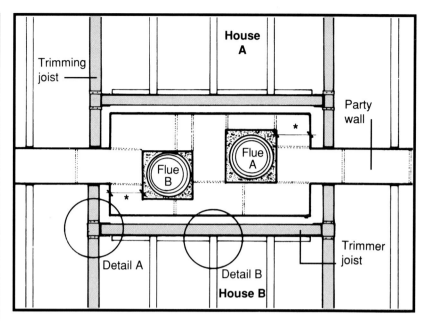

Joists perpendicular to chimney stack.

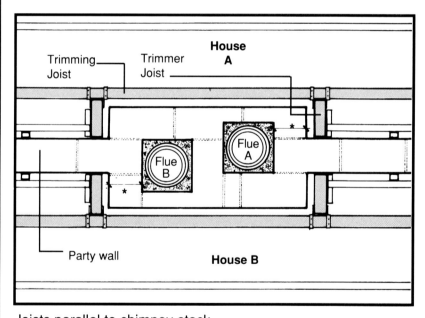

Joists parallel to chimney stack.

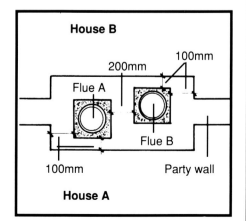

Minimum thickness for back to back chimneys.

Note:

(1) Joists are not built into the party wall – they are supported on hangers.

(2) The chimney breast dimensions illustrated on this page should be maintained up to the level of the roof covering. Above that, the 200mm dimension between flues and around flues can reduce to 100mm.

(3) * This dimension is illustrated as 200mm for the purpose of facilitating blockwork coursing and setting out. However it may actually be reduced to 100mm.

**TRIMMING AT FIRST FLOOR
LEVEL**
continued.

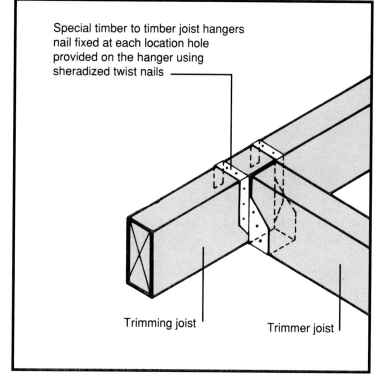

Special timber to timber joist hangers
nail fixed at each location hole
provided on the hanger using
sheradized twist nails

Trimming joist

Trimmer joist

Detail A

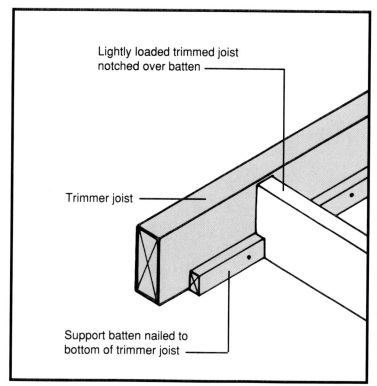

Lightly loaded trimmed joist
notched over batten

Trimmer joist

Support batten nailed to
bottom of trimmer joist

Detail B

HEIGHT OF CHIMNEY

For a chimney to draw properly it should extend a minimum distance above roof level. The distance varies with the pitch of the roof and the location of the stack on the roof relative to the ridge.

The sketch and table on this page indicate the minimum heights that chimneys should extend above roof level, as set out in Building Regulations Technical Guidance Document J. Local features such as sloping ground, tall trees or tall adjoining buildings may require the stack to be carried higher than the minimum dimensions shown here.

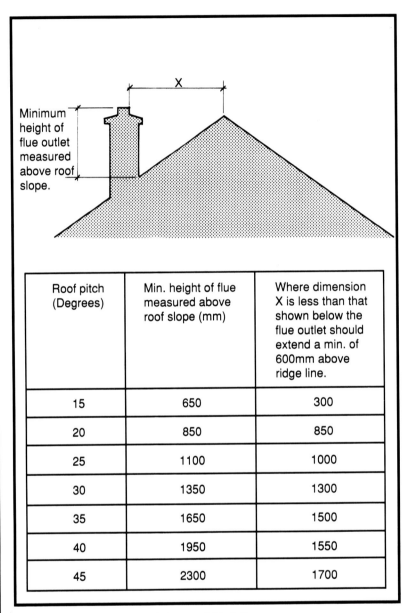

Roof pitch (Degrees)	Min. height of flue measured above roof slope (mm)	Where dimension X is less than that shown below the flue outlet should extend a min. of 600mm above ridge line.
15	650	300
20	850	850
25	1100	1000
30	1350	1300
35	1650	1500
40	1950	1550
45	2300	1700

Chimney height – provisions of Building Regulations Technical Guidance Document J.

IMPORTANT
A chimney that is not raised above ridge level may not draw properly. HomeBond recommends that the top of all stacks should extend above ridge level. Stability requirements should also be observed - see page 108.

HEIGHT OF CHIMNEYS
continued.

Flat roof extensions

The guidance on this page is in accordance with Technical Guidance Document J of the Building Regulations (as indicated by the shaded area) and HomeBond recommendations (as indicated by the broken lines).

In order for the chimney to draw properly it is recommended that stacks of the type illustrated should extend to ridge level as indicated by the broken line.

The height / width relationship of the chimney should satisfy the guidance on page 108.

A chimney must be built so that it has stability. It is obvious that a tall thin chimney could suffer storm damage. When placing chimney foundations ensure that they are of sufficient size to suit the chimney.

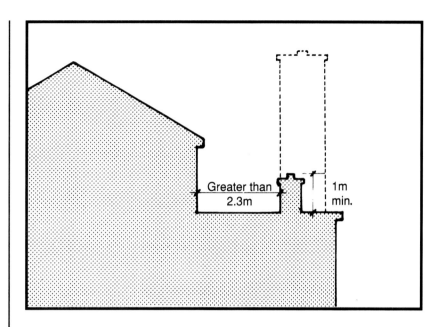

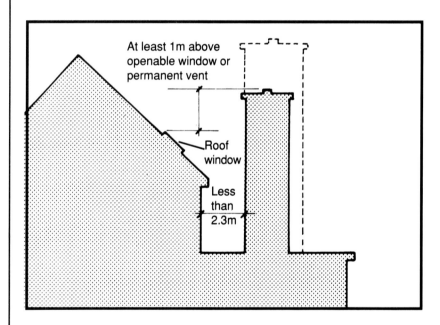

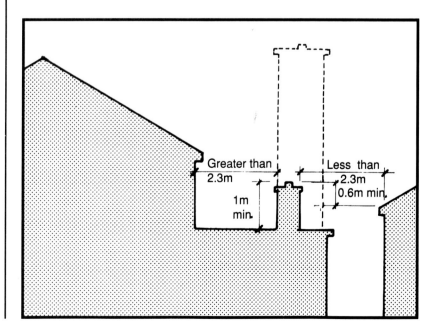

HEIGHT AND WIDTH OF CHIMNEY

Excessively slender chimneys may be liable to overturning in high winds. Chimneys are not usually designed for the fixing of aerials or satellite dishes.

The height / width provisions illustrated here are those set out in the Building Regulations Technical Guidance Document A.
In certain circumstances having regard to location and wind speeds these proportions may be modified, if the chimney design can be shown by calculation to be stable in the particular wind environment of the building.

For example:

1 For a standard stack having dimension (B) 550mm. The height (H) (measured as illustrated) should not exceed 1925mm.

2 Where the height (H) of a stack (measured as illustrated) is 3.5m the width (B) should be at least 1m.

Rule of thumb:
Where chimney height exceeds 2m, the width of the chimney will require increasing from standard width.

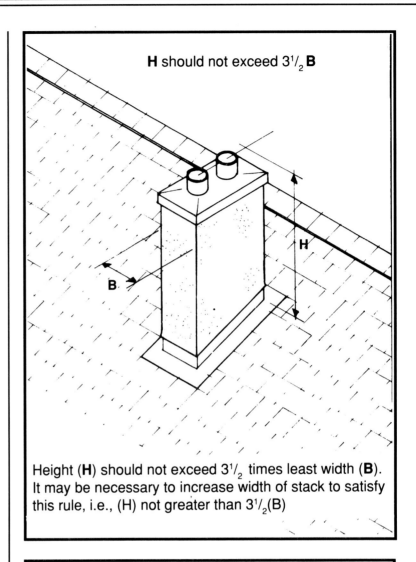

H should not exceed $3\frac{1}{2}$ **B**

Height (**H**) should not exceed $3\frac{1}{2}$ times least width (**B**). It may be necessary to increase width of stack to satisfy this rule, i.e., (H) not greater than $3\frac{1}{2}$(B)

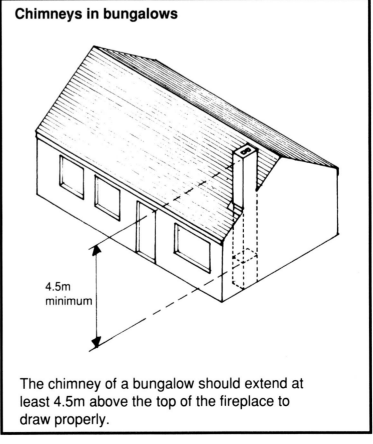

Chimneys in bungalows

4.5m minimum

The chimney of a bungalow should extend at least 4.5m above the top of the fireplace to draw properly.

HEIGHT AND WIDTH OF CHIMNEY
continued.

The outlet of a chimney or flue pipe must not be less than 1m above the top of any opening skylight (roof window) opening window or wall ventilator within 2.3m as illustrated.

Chimney cappings.

Take care with the construction of chimney cappings to avoid rain penetration into the top of the stack. The incorporation of a D.P.C. under the chimney capping is necessary.

Illustrated here are typical details of precast and in-situ concrete chimney cappings. The use of the precast option is recommended.

The chimney that will give least trouble from a rain penetration point of view is the rendered stack with a precast concrete chimney capping.

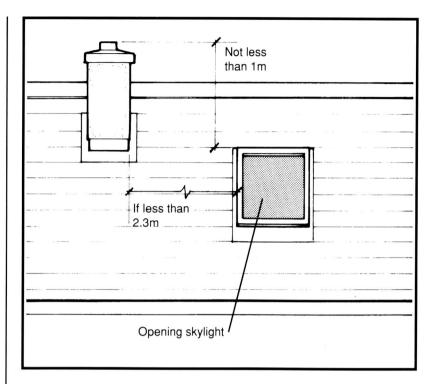

Precast concrete chimney capping.

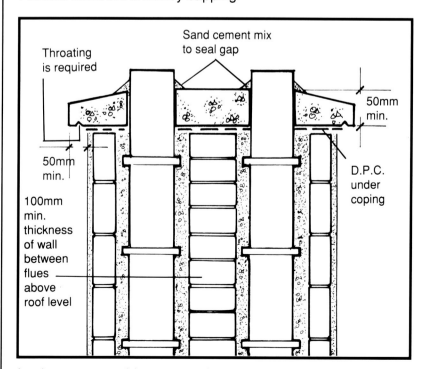

In-situ concrete chimney capping.

BRICK CHIMNEY STACKS

In brickwork chimney stacks, build in a metal tray d.p.c. across full width of chimney to protect against damp penetration through stack.

Metal tray d.p.c. should have adequate durability

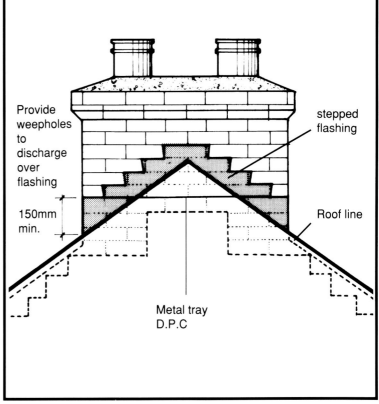

Location of metal tray D.P.C. in brickwork chimney stack.

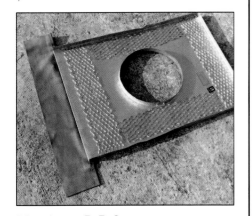

Metal tray D.P.C.

Properly installed metal tray D.P.C. in brickwork chimney. Wall separating flues on a party wall to be min. 200mm thick below roof level.

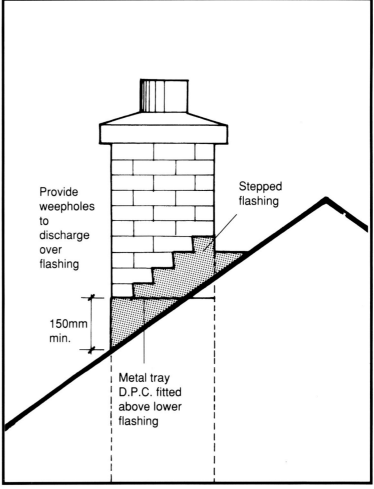

Location of metal tray D.P.C. in brickwork chimney stack.

AVOIDING DEFECTIVE AND SMOKING CHIMNEYS

The table on this and the following page highlights the ten most common faults in fireplaces and chimney construction and the steps that should be taken to avoid them.
Repairs to chimneys are costly.

See appendix C for recommended levels of ventilation for rooms in dwellings. Technical Guidance Document F of the Building Regulations advises that all habitable rooms must have permanent ventilation openings of not less than 6500mm². This includes rooms with a fireplace.

Defects	Causes	Prevention
1 Blockages	Mis-aligned flue lintels and liners. Mortar joints protruding into flue. Debris in flue.	Care in workmanship and supervision. Proper packing of flue liners as work proceeds. Clean and test flue thoroughly on completion of construction.
2 Air starvation	House too well sealed Note: Many builders fit underfloor air feed pipes to sunken ash box type fires to avoid air starvation problems	Ensure adequate air supply. If there is insufficient air supply, air speed through the flue will be inadequate for proper draught. If opening a door or window on a calm day stops the fireplace smoking, this usually indicates air starvation. Form throat properly – the throat should be approximately 300mm wide by 100mm. These are the smallest dimensions that will allow a chimney brush to pass through.
3 Badly-formed throat	Poor workmanship / supervision	Care in workmanship / supervision.Use a precast flue-gathering lintel to ensure smooth flow to flue
4 Steady downdraught	Top of chimney too low, e.g. not above ridge. Chimney too short (e.g. in bungalow)	Carry chimney above ridge level in all cases. Ensure that stack extends at least 4.5m above the top of the fireplace
5 Intermittent downdraught	Chimney near higher building or trees. Chimney located on a sheltered hillside or in a valley	Where practical, build chimney to a level where down draught is overcome. Where this is not practical, incorporate a draught-inducing cowl into the chimney construction (See illustrations on page 113).

AVOIDING DEFECTIVE AND SMOKING CHIMNEYS
continued.

Defects	Causes	Prevention
6 Fireplace opening too high	Opening such that top is more than 550mm above level of grate	Care in setting-out opening. Care in workmanship and supervision. Use standard sizes for fire openings
7 Badly-built offset	Offset too long, too shallow or too close to throat	Care in workmanship and supervision. Angle of offset should be no shallower than 52.5° to the horizontal. The straighter the chimney, the better the draw
8 Wrong size flue	Too big – chimney does not warm up. Too small – not permitted due to fire risk	Use standard flue liners of 200mm internal diameter for open fines
9 Air leaks	Open joints in blockwork. Flue liners not properly packed. Boiler pipes not sealed where they leave chimney	Seal joints in flue liners with mortar. Pack flue liners fully with weak wetted mortar mix having cement content. Fill all joints in blockwork to chimney. Apply coat of sand / cement render to chimney blockwork at ground floor level and at first floor level. Pay particular attention at ceiling/floor joist level to ensure all joints adequately sealed. Seal thoroughly primary and secondary circuit pipes from back boilers and room heaters where they pass through the side of the chimney. It is a fallacy that a chimney with bends draws better than a straight flue. The straighter the chimney is, the better it draws. Never fix dry lining to a chimney breast unless the breast has been first rendered in sand/cement. An unrendered breast can allow smoke to leak from the fire and enter habitable areas. Any drylining fixed to a chimney breast must be fully sealed at mantel and ceiling level in the room containing the fireplace.
10 Unsuitable chimney pot	Pot smaller than flue or tapering to a diameter less than of the flue liner	Use a standard flue liner as the flue terminal, or alternatively use a terminal having a minimum diameter at least equal that of the flue liner.

AVOIDING DEFECTIVE AND SMOKING CHIMNEYS
continued.

As stated in item (5) of the preceding table, the incorporation of a chimney cowl may help in overcoming downdraught in chimneys. The sketches on this page illustrate typical purpose-made cowls which can be used in such circumstances. Proper design and construction of the chimney should overcome the need to have recourse to such devices in the majority of cases.

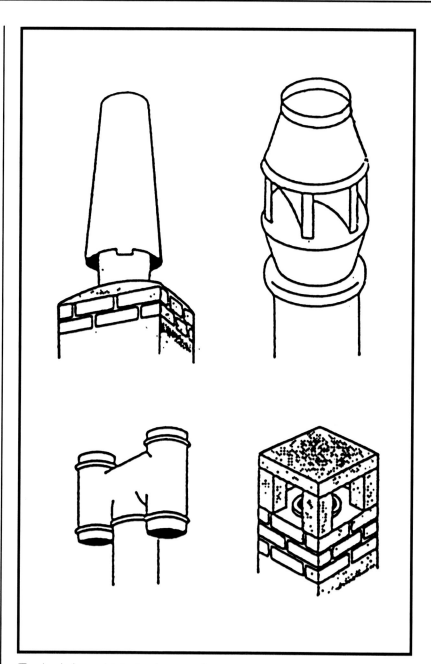

Typical draught-inducing cowls.

INTRODUCTION

SR11: "Structural Timber for Domestic Construction", is the Irish Standard Recommendation that must be used for structural timber applications. The document is designed to be used for sizing kiln dried timber with maximum moisture content of 22% for normal domestic houses up to three storeys.

Use the tables in SR11 to ensure correctly sized joists and satisfactory performance.

For the design of structural timber outside the scope of SR11 an engineer should be engaged. The engineer appointed must be qualified by examination, be in private practice and possess professional indemnity insurance.

The following pages 114 to 117 explain SR11, give worked examples and reproduce the tables.

SR11 and an explanatory leaflet is available from:
EOLAS,
The Science and Technology Agency,
Glasnevin,
Dublin 9.

STRENGTH CLASSES

SR11 divides timber in ascending order of strength into three strength classes (SC A, SC B and SC C) depending on the species and grade of timber.

The particular species and grades of Irish and imported timbers that fall into these Strength Classes are set out in the table below.

Soft wood Species	Strength classes		
	SC A	SC B	Sc C
Irish Timber:			
Sitka Spruce	GS	SS	M75
Norway Spruce	GS	SS	M75
Douglas Fir	GS		SS
Larch		GS	SS
Imported Timber:			
Whitewood*		GS	SS
Redwood*		GS	SS
Fir-Larch**		GS	SS
Spruce-Pine-Fir**		GS	SS
Hem-fir**		GS	SS

* European ** North American

Timber Grades
The grade abbreviations in the above table are as follows:

Visually graded
General Structural: GS
Special Structural: SS

Machine grade:
M75

INTRODUCTION
continued.

It is a requirement of SR11 that all structural timber used shall be stress graded and marked accordingly. The marking system identifies the stress grade and Strength Class of the timber and the registered number of the timber grader and the company. Both the grading and marking of timber by individual companies is subjected to the supervisory control of:

The Timber Quality Bureau of Ireland Forest Products Department, EOLAS.

On this page are illustrated examples of the markings which occur on stress graded timber in accordance with SR11.

Span tables
The following page contains the span table for floor joists from SR11 and examples of its application. The key factor when reading the span tables is to correlate the span, section size and the joist centres applicable with the Strength Class of the timbers. It will be noted that the permissible spans increase from SC A to SC B to SC C for the same section sizes and spacings.

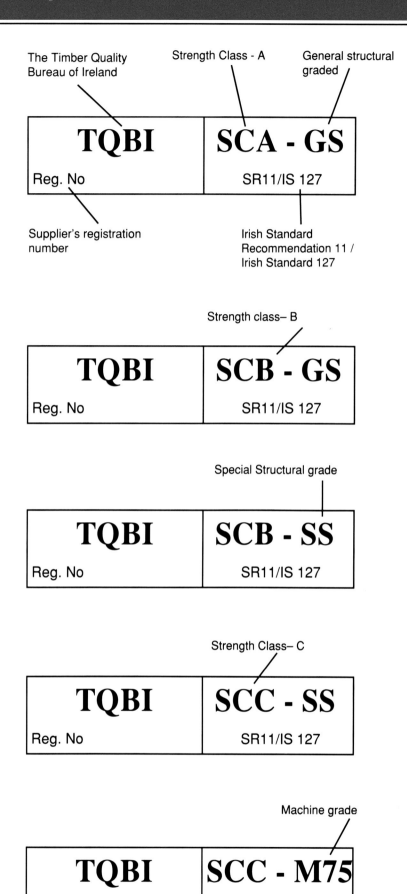

The Timber Quality Bureau of Ireland

Strength Class - A

General structural graded

TQBI	SCA - GS
Reg. No	SR11/IS 127

Supplier's registration number

Irish Standard Recommendation 11 / Irish Standard 127

Strength class– B

TQBI	SCB - GS
Reg. No	SR11/IS 127

Special Structural grade

TQBI	SCB - SS
Reg. No	SR11/IS 127

Strength Class– C

TQBI	SCC - SS
Reg. No	SR11/IS 127

Machine grade

TQBI	SCC - M75
Reg. No	SR11/IS 127

SIZES OF FLOOR JOISTS

Size of joist (mm)	Strength Class								
	SC A			SC B			SC C		
	Spacing of joists								
	300	400	600	300	400	600	300	400	600
	Permissible span of joists in metres								
35 x 100	2.02	1.81	1.48	2.12	1.92	1.67	2.21	2.00	1.74
35 x 115	2.33	2.07	1.69	2.43	2.21	1.92	2.54	2.30	2.00
35 x 125	2.54	2.24	1.83	2.65	2.40	2.09	2.76	2.50	2.18
35 x 150	3.04	2.66	2.17	3.18	2.88	2.51	3.31	3.01	2.62
35 x 175	3.54	3.08	2.51	3.71	3.37	2.93	3.87	3.51	3.05
35 x 200†	4.04	3.50	2.85	4.25	3.85	3.34	4.42	4.01	3.49
35 x 225†	4.51	3.91	3.19	4.78	4.33	3.73	4.98	4.51	3.93
44 x 100	2.19	1.98	1.66	2.29	2.08	1.81	2.39	2.16	1.89
44 x 115	2.52	2.28	1.89	2.63	2.39	2.07	2.74	2.48	2.16
44 x 125	2.74	2.48	2.05	2.87	2.59	2.25	2.98	2.71	2.35
44 x 150	3.28	2.98	2.43	3.44	3.12	2.72	3.58	3.25	2.83
44 x 175	3.84	3.45	2.82	4.02	3.65	3.16	4.18	3.79	3.30
44 x 200	4.38	3.92	3.20	4.59	4.16	3.62	4.78	4.33	3.77
44 x 225†	4.94	4.38	3.57	5.17	4.69	4.08	5.38	4.88	4.24
63 x 150	3.71	3.37	2.91	3.89	3.52	3.07	4.04	3.64	3.20
63 x 175	4.33	3.93	3.37	4.54	4.12	3.58	4.72	4.28	3.73
63 x 225	5.58	5.06	4.28	5.83	5.29	4.61	6.07	5.51	4.80
75 x 150	3.94	3.58	3.11	4.12	3.74	3.26	4.29	3.89	3.40
75 x 175	4.60	4.17	3.63	4.81	4.37	3.80	5.01	4.54	3.96
75 x 225	5.91	5.37	4.67	6.19	5.61	4.89	6.44	5.84	5.09

* The above joist sizes are the minimum permissible sizes at 22% moisture content.

** The permissible span is the clear span between supports.

† This joist requires bridging at intervals of 1350 mm.

TABLE 5 : SR11. : FOR USE IN SIZING DOMESTIC FLOOR JOISTS. THE USE OF THIS TABLE IS ESSENTIAL TO ENSURE PROPERLY SIZED AND SPACED FLOOR JOISTS. NOTE: THIS TABLE DOES NOT ALLOW FOR POINT LOADS SUCH AS PARTITIONS.

SIZING FLOOR JOISTS

The following example illustrates the use of the span tables in sizing floor joists.

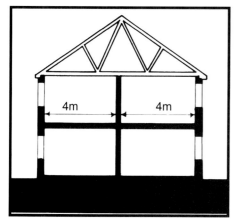

A house with an internal span of 8m has an internal load bearing wall provided at mid-span as illustrated. The timber specified is Irish grown Sitka Spruce of General Structural Grade (GS).

From the table on page 115 this type of timber is in Strength Class A (SC A).

From the extract of the span tables reproduced below a floor joist spanning 4m can be chosen from either 35 x 200mm at 300mm centres, or 44 x 225 at 400 centres (Note: These joists require bridging at 1350mm centres).

Sizes of floor joists			
	Strength Class		
	SC A		
Size of joist (mm)	Spacing of Joists (mm)		
	300	400	600
	Permissible Span in (M)		
35 x 100	2.02	1.81	1.48
35 x 115	2.33	2.07	1.69
35 x 125	2.54	2.24	1.83
35 x 150	3.04	2.66	2.17
35 x 175	3.54	3.08	2.51
35 x 200†	4.04	3.50	2.85
35 x 225†	4.51	3.91	3.19
44 x 100	2.19	1.98	1.66
44 x 115	2.52	2.28	1.89
44 x 125	2.74	2.48	2.05
44 x 150	3.28	2.98	2.43
44 x 175	3.84	3.45	2.82
44 x 200	4.38	3.92	3.20
44 x 225†	4.94	4.38	3.57
63 x 150	3.71	3.37	2.91
63 x 175	4.33	3.93	3.37
63 x 225	5.58	5.06	4.28
75 x 150	3.94	3.58	3.11
75 x 175	4.60	4.17	3.63
75 x 225	5.91	5.37	4.67

† This joist requires bridging at intervals of 1350mm.

Depending on the joist span, availability and cost of material, the most economical section can be selected. This also applies to choice of Strength Class provided that this is not specified.

FLOOR JOISTS ADJOINING EXTERNAL WALLS

To help stabilise external walls it is necessary to tie the walls to the floor.

Technical Guidance Document A of the Building Regulations recommends such strapping for joists adjoining external walls including gable walls.

Strapping of floor joists to external wall – Galvanised steel straps 30 x 5mm min. in cross section carried over three joists and built into the external wall as work progresses. Straps to be provided at 2m max. centres. Where continuity is disrupted by openings such as stairwells, such openings should not exceed 3m in length, and the distance between centres of straps on either side of the opening should be reduced to compensate for any omitted along the length of the opening. Provide nogging pieces between joists directly below straps, together with packing pieces between joists and wall, as illustrated.

Straps need not be provided:

(a) In the longitudinal direction of joists in houses of not more than two storeys, if the joists are at not more than 1.2m centres and have at least 90mm bearing on the supported walls or 75mm bearing on a timber wall plate at each end,
or

(b) In the longitudinal direction of joists in houses of not more than 2 storeys, if the joists are carried on the supported wall by joist hangers of the restraint type described in I.S. 325 Part 1:1986.

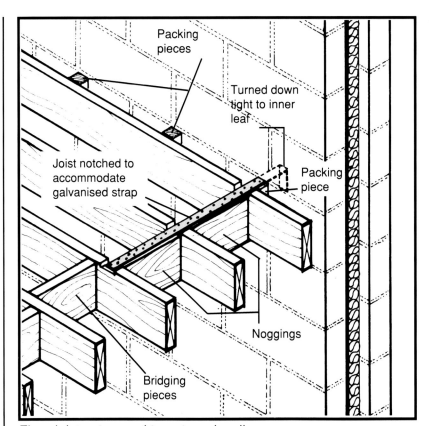

Floor joists strapped to external wall.

TRIMMING AROUND OPENINGS

Stairwell opes:

Trimmers and trimming joists to stairwell opes not exceeding 3m in length and 1m in width should be sized in accordance with the table opposite.

For stairwell opes exceeding 3 x 1m trimmer and trimming joists may require specialist design by an engineer. The engineer appointed should be qualified by examination, be in private practice and possess professional indemnity insurance.

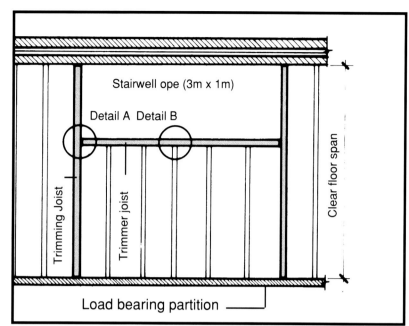

Trimming around stairwell ope
(see page 105 for details A and B).

Size of trimmer/ trimming joist (mm)	Permissible clear floor span (m)		
	SC A	SC B	SC C
75 x 225	2.6	3.1	3.4
2 no. 44 x 225*	3.0	3.5	3.8
75 x 175	2.0	2.4	2.5
2 no. 44 x 175*	2.3	2.7	2.8

Maximum permissible clear floor spans for different trimmer/trimming joist sizes and their strength classes.

* If using a pair of joists to form trimming/trimmer joists. They should be nailed at 450mm centres 20mm from top and bottom edges of the joist. Alternatively they can be bolted on the centreline at approx. 1m centres. If using this method it is difficult to obtain joist hangers to accommodate the uncommon width of two joists together.

ENSURING RIGID FLOORS

To generally stiffen the floor and to reduce the risk of joist twisting it is essential to provide adequate bridging. Bridging should be solid timber the same depth as the joist or, in limited situations, of the herring–bone variety. Frequency of bridging should generally be as follows:

Span (m) of floor joists	Number/position of rows of bridging timbers
Up to 2.5	None required
2.5 to 4.5	One row at mid–span
Over 4.5	Two rows – one at each third–of–span position
In addition, any 35 x 200mm 35 x 225mm and 44 x 225mm joists should be bridged at 1350mm centres.	

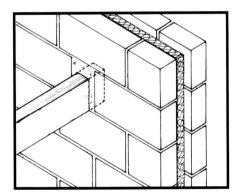

Where joists are built into walls ensure a min. bearing of 90mm, and the space between the joists and the blockwork should be packed solidly with mortar to ensure rigidity as illustrated above.

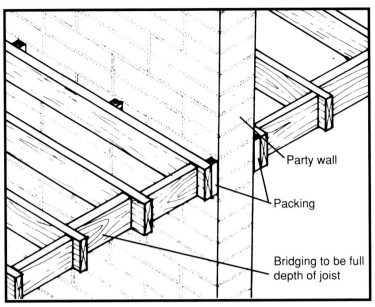

Provide packing and bridging.

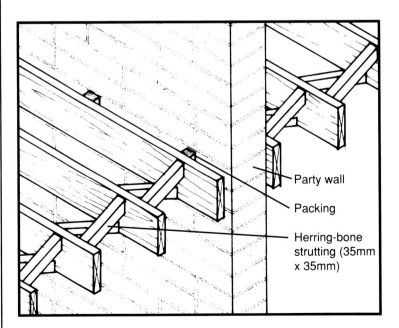

Alternative to the above: Herring–bone strutting and packing. Such strutting is only acceptable where joist spacing does not exceed three times the joist depth. For example, at 400mm centres, only joists up to 125mm in depth may have herring–bone struts. Above that depth, solid bridging should be used.

NOTCHING AND DRILLING

Extreme care should be taken when notching and drilling joists. They should only be notched and drilled in the location shown and to the extent shown.

NOTCHES SHOULD ONLY BE MADE IN THE TOP OF THE JOIST

Care must be exercised to ensure that the depth of saw-cut for notching does not extend beyond the depth of the notch itself (i.e. one eighth of the joist depth, maximum).

Note:
Make clear to plumbers and electricians, that they must not weaken joist by inappropriate notching and drilling.

The horizontal distance between any hole and any notch should not be less than the depth of the joist.

NOTCHING: DO NOT NOTCH IN THE SHADED AREA

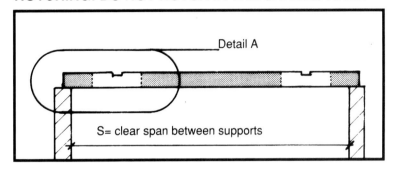

S= clear span between supports

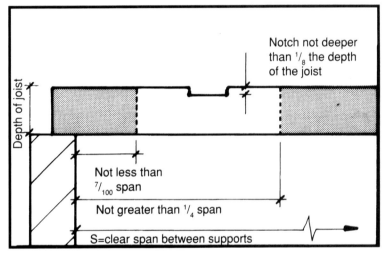

Detail A

DRILLING: Do not drill in the shaded areas

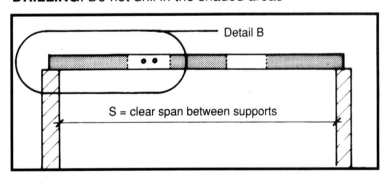

S = clear span between supports

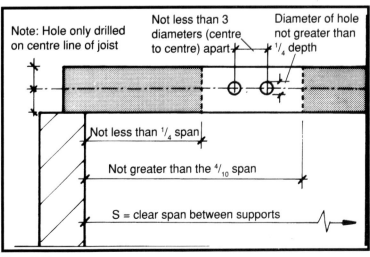

Detail B

FLOORING MATERIALS

The table opposite outlines the maximum spans for different thicknesses of flooring material.

Flooring material	Thickness mm	Max. joist centres mm
t & g softwood boards	16	500
	19	600
Flooring – grade Chipboard***	18	500
	22	600
Finply t & g Plywood	12	400*
	15	450**
Canadian Fir t & g Plywood	12	400
	15	600

* Joist centres may increase to 450mm if joints at right angles to joists are supported on noggings.

** Joists centres may increase to 600mm if joints at right angles to joists are supported on noggings.

***The use of chipboard in wet areas such as bathrooms is not recommended.

STUD PARTITIONS

A typical timber stud partition consists of a layer of plasterboard nailed to each side of a framework of timber.

The plasterboard for this partition is supplied with tapered edges for flush jointing.

It is essential that the timber framing be accurately spaced and aligned. The timber used should be straight and properly dried (max. 22% moisture content) so as to prevent warping or distortion, which can cause cracking of the joints.

Non-loadbearing stud partitions (such as those used at first floor level in typical two-storey housing) should be of kiln dried timber and comprise of 35mm x 75mm minimum studs at 400 mm centres.

Load-bearing stud partitions, when used, should comprise timber of at least Strength Class A and should incorporate studs of 44 x 100mm minimum at 400mm centres with double header and sole pieces. Load-bearing partitions may require fire resistance if carrying floors (see appendix J, Fire).

All stud partitions should include two rows of noggings as illustrated on this page, each end of the nogging to be double nailed. If the boards are placed horizontally across the studs, noggings are required behind the horizontal joints.

The plasterboard should be nailed not closer than 12.7mm to the edges and the nails should be driven straight (i.e. not skewed) to a secure bed in the timber. When two cut edges are to be fixed to a timber support, they should be tightly butted together. A 3mm gap should be left between the bound edges of plasterboard.

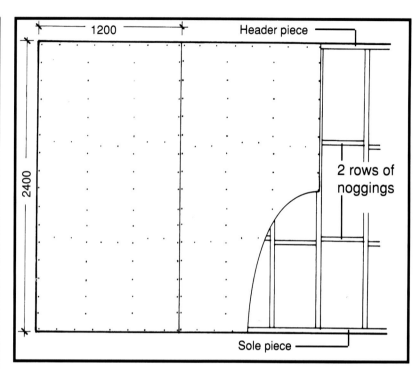

Typical stud partition.
(Non-loadbearing)

Supporting members	Board thickness (mm)	Board width (mm)	Recommended Max. centres (mm)
Studs	9.5	900	450
		1200	400
	12.5	900	450
		1200	600

The table above sets out the recommended maximum spacing of studs for the range of standard plasterboard thicknesses and sheet widths.

FIXING DETAILS

Stud partition construction: typical fixing details.

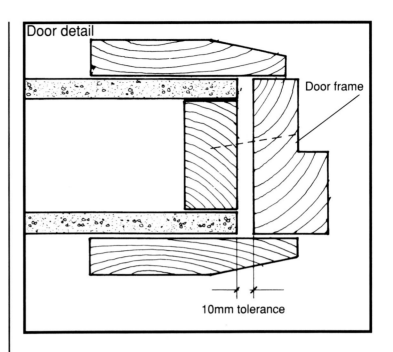

Door detail

Door frame

10mm tolerance

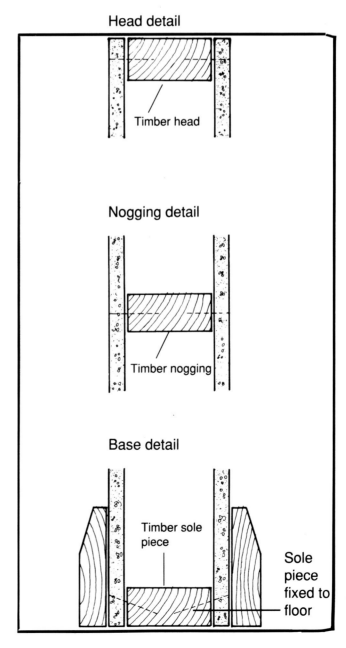

Head detail

Timber head

Nogging detail

Timber nogging

Base detail

Timber sole piece

Sole piece fixed to floor

CONSTRUCTION

Non–loadbearing stud partitions should be built off double joists to avoid the risk of ceiling deflection and cracking, or, alternatively, above an extra joist placed directly under the line of the partition as shown.

Note:
If the partition is load bearing (e.g. carrying roof loads) a specially designed member such as an RSJ or universal beam may be required to support it.

DO NOT LOAD ROOF UNTIL PARTITIONS ARE PROPERLY SECURED INTO PLACE.

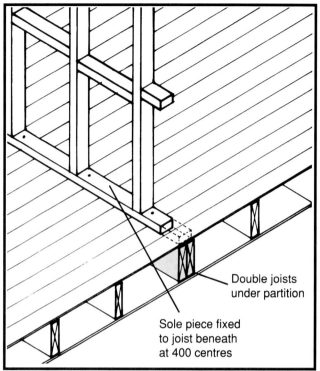

Double joists under partition

Sole piece fixed to joist beneath at 400 centres

Typical non–load bearing stud partition built off timber floor, with double joists beneath.

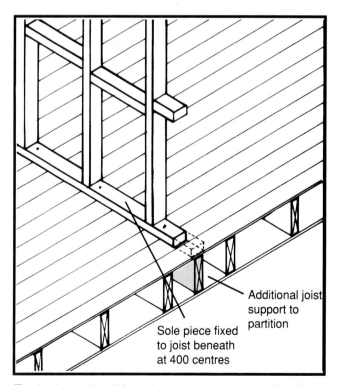

Additional joist support to partition

Sole piece fixed to joist beneath at 400 centres

Typical non-load bearing stud partition built off timber floor. Note the additional joist to support partition.

CONSTRUCTION
continued.

It should be noted that where partitions run at right angles to the joists that the design of the joists should take into account the point load imposed thereby. Any partition at right angles to the joist should incorporate a double sole piece.

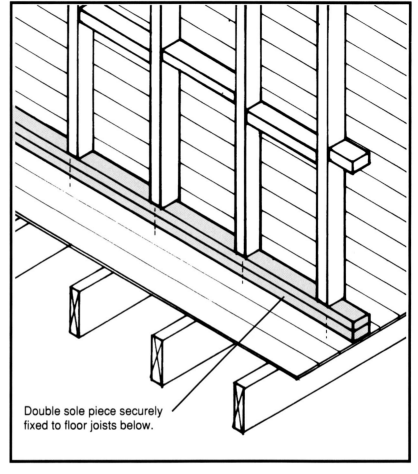

Double sole piece securely fixed to floor joists below.

Partition at right angles to joist – note that joist sizing should take partition load into account. Note also double sole piece.

The heads of partitions should be fixed to the ceiling joists if the partition is at right angles to the joists, or, if the partition is parallel to the joists it should be fixed to nogging placed between the joists. The noggings should be spaced at 400mm centres.

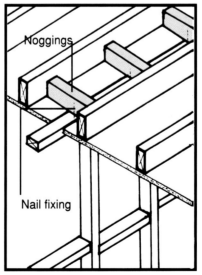

Noggings

Nail fixing

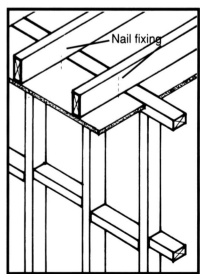

Nail fixing

Details at head of partition.

Note:
The sketches above illustrate stud partitions under a cut timber roof construction, where the ceiling material is fixed after the partition is erected. However with prefabricated trussed roof construction the ceiling material is often fixed before the partition is erected.

INTRODUCTION

A stairway may be constructed with steps rising without a break from floor to floor, or with steps rising to a landing between floors, with a further series of steps rising from the landing to the floor above. The most common type in domestic construction is the straight flight stair consisting of a straight flight of parallel steps.

Definitions.

"**flight**" is a part of a stairway which consists of a step or consecutive steps;

"**going**" means the horizontal distance between the nosing of a tread and the nosing of the tread or landing next above it;

"**pitch**" means the angle between the pitch line and the horizontal;

"**rise**" means the vertical distance between the top of a tread and the top of a tread, landing or ramp next above or below it;

"**tread**" means the upper surface of a step within the width of a stairway;

"**string**" a sloping board at each end of the treads housed or cut to carry the treads and risers of a stair. A string can be either a wall string or an outer string and either a closed string or a cut string.

Note: In any single flight of stairs the rise and going of each step must be consistent.

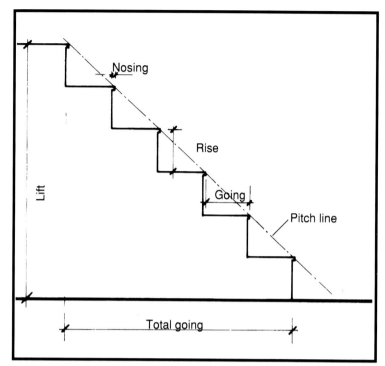

Straight flight stairs.

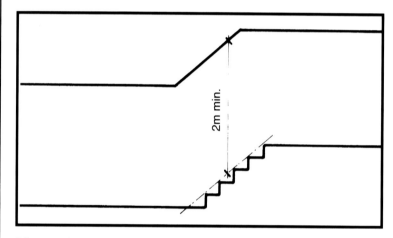

Measuring headroom.
Headroom over the whole width of any stairway measured as shown should be a minimum of 2m. measured vertically above the pitch line.

Relationship between rise and going.

In order to ensure steps are suitably proportioned the sum of twice the rise plus the going (2R +G) should be between 550mm and 700mm.

Dimensions for going: 250mm max. (220mm optimum).

Dimensions for Rise: 220mm max. (175mm optimum).

Angle of pitch: 42° max. (35° optimum).

HANDRAILS

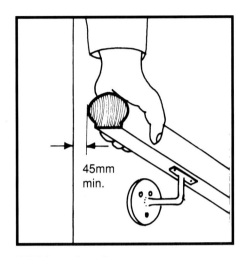

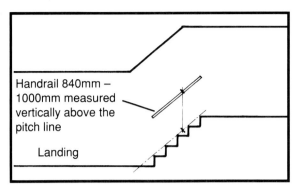

Handrail 840mm – 1000mm measured vertically above the pitch line

Landing

Widths of stairs.

Technical Guidance Document K of the Building Regulations recommends a minimum unobstructed width of 800mm for domestic stairs. In the case of flats/apartments the minimum unobstructed width should be 1000mm.

Landings

A landing should be provided at the top and bottom of every flight except that a landing may not be necessary between a flight and a door if the total rise of the flight is not more than 600mm and the door slides or opens away from the stairs. The landing may include part of a floor. The width and going of the landing should be at least as great as the smallest width of the stairway.

Guarding.

Stairways should be guarded at the sides (where the total rise of any flight is no more than 600mm, guarding may not be essential). Suitable guarding would include a wall, screen, railing or balustrade.

Height of guarding.

For domestic stairways, guarding height should be 840mm minimum measured vertically above the pitchline on flights and 900mm minimum on landings.

Handrails.

A stairway should have a handrail on at least one side if it is less than 1000mm wide, and should be designed to ensure: **(1)** A firm hand hold. **(2)** That trapping or injuring of the hand is prevented. **(3)** A minimum 45mm clearance at the back of the handrail. **(4)** Secure fixing. **(5)** That handrail ends do not project to catch clothing etc. **(6)** The handrail ends should allow uninterrupted hand freedom along the length of the flight.

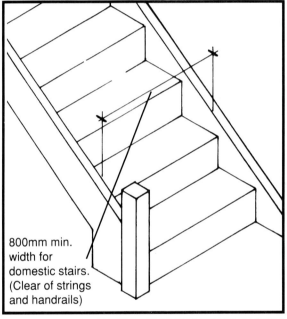

800mm min. width for domestic stairs. (Clear of strings and handrails)

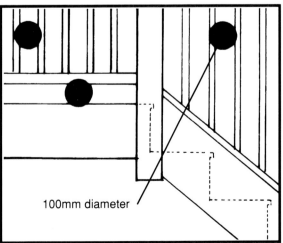

100mm diameter

Guarding balusters should be constructed so that a 100mm diameter sphere cannot pass through any openings in the guarding and so that it will not be readily climbable by children.

TIMBER STAIRCASE CONSTRUCTION

String: The string is the inclined member which supports the ends of treads and risers.

Closed string: (as illustrated) an outer string with parallel edges into which the treads and risers are housed.

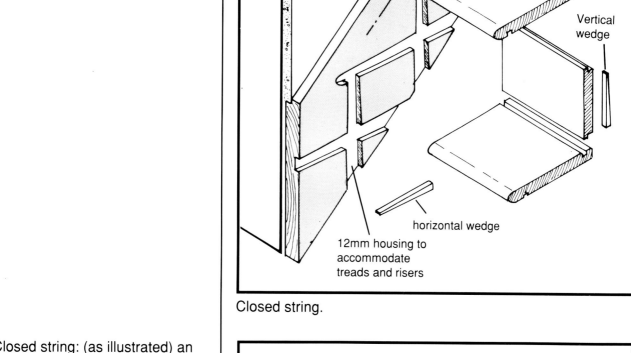

Screw fix treads to risers at 240mm centres

50mm margin

Vertical wedge

horizontal wedge

12mm housing to accommodate treads and risers

Closed string.

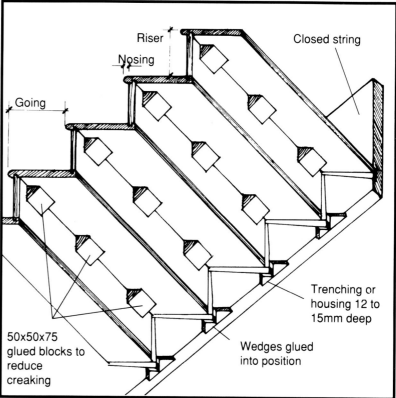

Riser

Nosing

Closed string

Going

Trenching or housing 12 to 15mm deep

50x50x75 glued blocks to reduce creaking

Wedges glued into position

Underside of typical staircase.

GUARDING

Guarding should be provided to the sides of any part of a raised floor, gallery, balcony or roof or any other place to which people have access (unless access is only for the purpose of maintenance or repair).

Guarding requirements:

◆ Suitable guarding would include a screen, railing, or balustrade and should be a minimum of 900mm high above floor level and should be capable of resisting a horizontal force of 0.36kN/m applied at a height of 900mm above floor level.

◆ Appropriate precautions should be taken where the cill of a window is below 900mm.

◆ Guarding should be designed and built in such a way that it does not present unacceptable risks of accidents in service.

◆ Guarding should be so constructed that a 100mm diameter sphere cannot pass through any openings in the guarding.

◆ Guarding should not be readily climbable by children.

This type of horizontal guarding is not acceptable as it can be readily climbed by children.

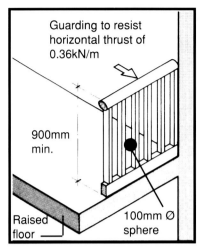

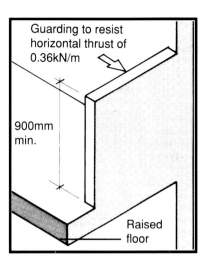

Balustrade or railing type guarding should be constructed so that a 100mm dia. sphere cannot pass through the opening in the guarding.

Solid guarding.

Note: Guarding to balconies should be provided at a height of at least 1100 mm above balcony level.

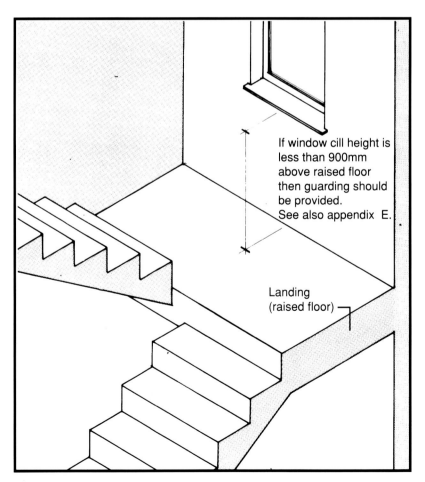

If window cill height is less than 900mm above raised floor then guarding should be provided. See also appendix E.

Landing (raised floor)

Window Cills.
Note: In the interest of clarity, guarding for the stairs has been omitted from this sketch.

ROOF CONSTRUCTION

INTRODUCTION

The fundamental requirements of a properly constructed roof apart from prevention of moisture penetration, are that it should be built to prevent excessive deflection and to properly transfer the roof loads to the load bearing walls of the house.

The underlying principles of roof construction are relatively simple and they are outlined on the following pages for prefabricated trusses, traditional cut roofs, and flat roofs.

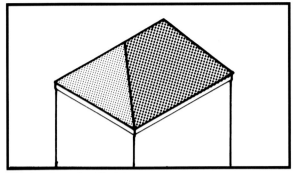

Pitched roof with hips

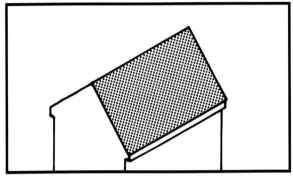

Standard pitched roof with gable end

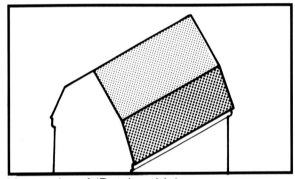

Mansard roof (Dutch gable)

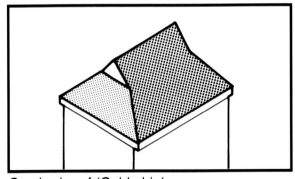

Gambrel roof (Gable hip)

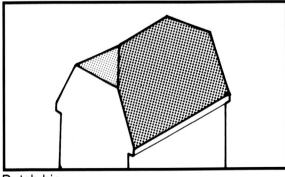

Dutch hip

INTRODUCTION

The vast majority of modern houses are built using prefabricated trussed rafters, usually placed at 600mm centres. These trusses rely on a number of factors for their successful performance, e.g,

◆ Proper design and fabrication by the manufacturer for the roof in question.
◆ Galvanised punched metal plate fixings at joints.
◆ Adequate bracing.
◆ Adequate provision for holding down.
◆ Proper distribution of point loads such as water tanks.

The appropriate standard for roof construction incorporating such trusses is 'Irish Standard 193: 1986: Timber Trussed Rafters for Roofs' published by the National Standards Authority of Ireland and available from EOLAS, the Irish Science and Technology Agency, Glasnevin, Dublin 9.

Prefabricated trusses must bear the Timber Quality Board of Ireland (TQBI) approval tag. A selection of these tags and an explanation of their markings, are illustrated opposite.

When ordering trusses, design drawings and details of the roof should be provided to the suppliers to facilitate design and manufacture of the trusses.

Note: When measuring the span of trusses, measure from the outside of the wall plate on one side to the outside of the wall plate on the other side, and add 25mm.

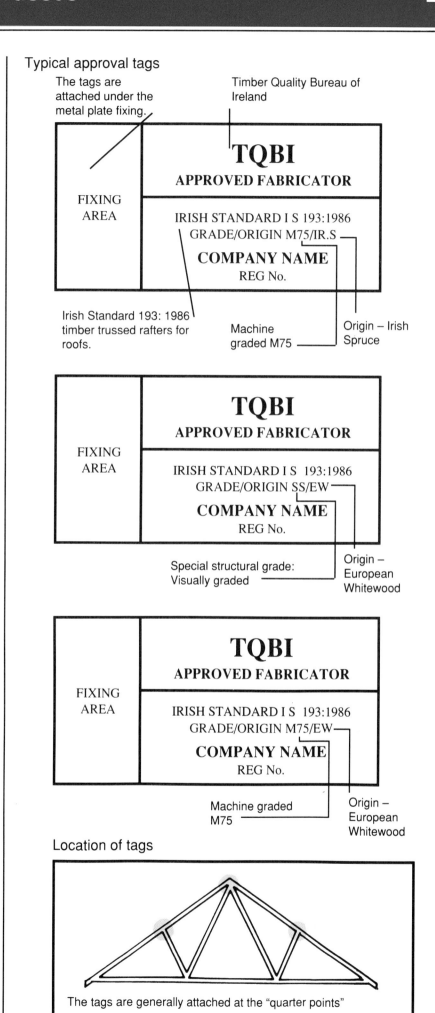

Typical approval tags

The tags are attached under the metal plate fixing.

Timber Quality Bureau of Ireland

FIXING AREA

TQBI
APPROVED FABRICATOR

IRISH STANDARD I S 193:1986
GRADE/ORIGIN M75/IR.S
COMPANY NAME
REG No.

Irish Standard 193: 1986 timber trussed rafters for roofs.

Machine graded M75

Origin – Irish Spruce

FIXING AREA

TQBI
APPROVED FABRICATOR

IRISH STANDARD I S 193:1986
GRADE/ORIGIN SS/EW
COMPANY NAME
REG No.

Special structural grade: Visually graded

Origin – European Whitewood

FIXING AREA

TQBI
APPROVED FABRICATOR

IRISH STANDARD I S 193:1986
GRADE/ORIGIN M75/EW
COMPANY NAME
REG No.

Machine graded M75

Origin – European Whitewood

Location of tags

The tags are generally attached at the "quarter points"

TRUSS BEARING

The truss must be supported directly on the ceiling chord directly below the point of intersection of joist and rafter, unless the trusses are specifically designed. Special design must be carried out in accordance with the approval of the truss manufacturer.

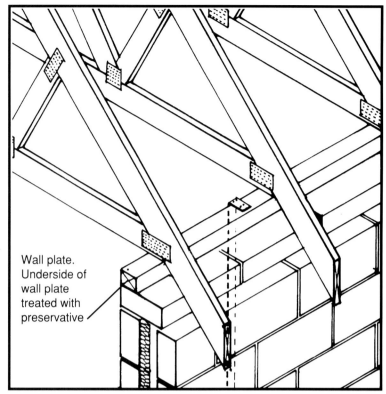

Wall plate.
Underside of wall plate treated with preservative

Truss bearing.

Overhangs of the type illustrated opposite are acceptable if the rafters and joists are braced in accordance with the manufacturers recommendations. **Note:** The truss shown opposite is a specially designed and manufactured cantilever truss.

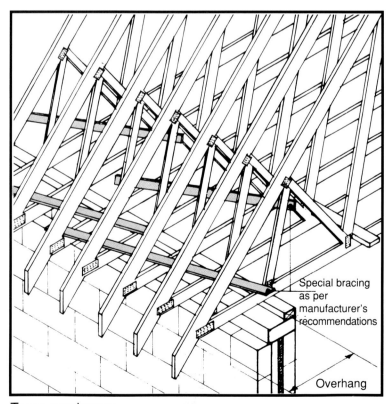

Special bracing as per manufacturer's recommendations

Overhang

Truss overhangs.

HOLDING DOWN METHODS

Trusses should be securely tied down to the walls as required to resist uplift by the wind. Fixing by proprietary clips or straps is the preferred method of fixing. Alternatively, fixing by means of nails may be considered, at least two wire round nails each not less than 4.5mm in diameter and not less than 100mm in length should be used to secure each truss to the wall plate in such a way as to prevent damage to the plate connectors. The nails should comply with I.S. 105, be staggered in position and be fixed on either side of the truss member.

DO NOT NOTCH RAFTERS TO SEAT TRUSSES ON WALL PLATE.

Truss clip should be used to fix each truss to the wall plate

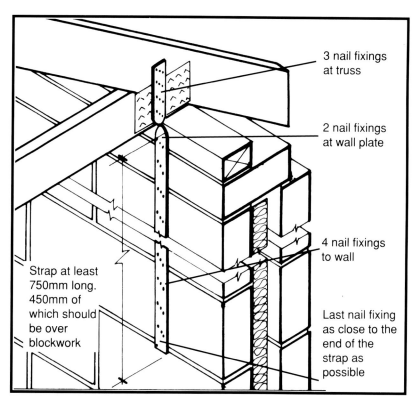

3 nail fixings at truss

2 nail fixings at wall plate

4 nail fixings to wall

Last nail fixing as close to the end of the strap as possible

Strap at least 750mm long. 450mm of which should be over blockwork

Twisted galvanised vertical restraint straps tying truss down to blockwork in wall. Min. cross section 30x5mm. Starting at truss nearest gable/ party wall and then every 2m centres max.

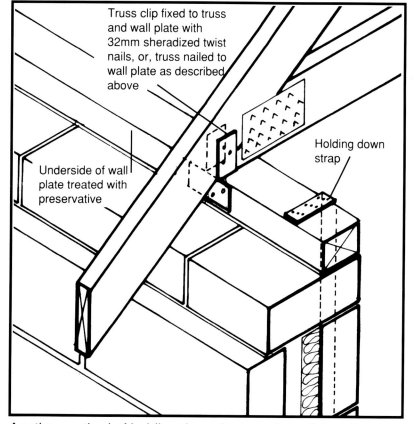

Truss clip fixed to truss and wall plate with 32mm sheradized twist nails, or, truss nailed to wall plate as described above

Holding down strap

Underside of wall plate treated with preservative

Another method of holding down the truss is by fixing directly to the wall plate. Wall plate should be secured by galvanised holding down straps, at 2m centres, min. cross section 30x5mm and carried down over at least 2 full block courses. Where wall plates are butt jointed holding down straps should be provided at not more than 400mm on either side of the joint.

BRACING

It is vital for the satisfactory performance of a prefabricated trussed roof that the trusses are tied together and braced properly so that the roof structure acts as a unit.

The minimum bracing details illustrated on this page are appropriate for the majority of roofs. Roofs of irregular or unusual design, or having a large span will require additional bracing which should be designed by an appropriately qualified person.

The following pages 136 to 138 illustrates some typical examples of such additional bracing.

Detailed design guidance on the bracing of trusses is given in I.S. 193 : 1986 (Timber Trussed Rafters for Roofs) published by the National Standards Authority of Ireland, Dublin 9.

Note:
Bracing details must be confirmed with the truss manufacturer / roof designer's recommendations.

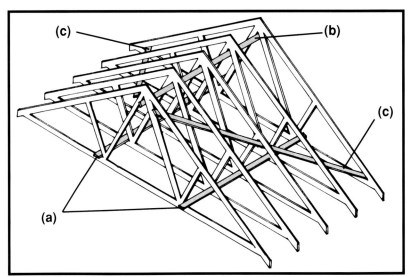

This sketch illustrates the minimum standard bracing of a prefabricated trussed roof.
(a) At ceiling level at intersection of ceiling joist, tie and strut i.e. node point.— These bracing timbers are called "longitudinal binders"
(b) At ridge level — these are also termed "longitudinal binders"
(c) From eaves to ridge on both sides of roof — this is called "rafter diagonal bracing".

Bracing timbers should be at least 100x25mm (nominal sizes) free from major defects and fixed with two 75mm galvanized nails of 2.65mm diameter at every point of contact with a truss. Where a single length of timber is not used lengths should overlap across two trusses.

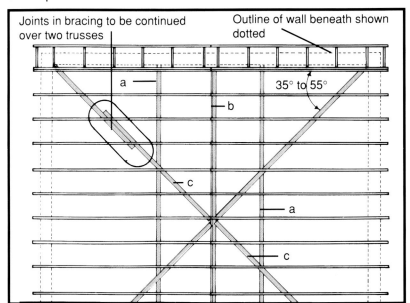

Rafter diagonal bracing should be fixed at an angle of approximately 35° to 55° on plan to the rafters and should repeat continuously along the roof. The end of all longitudinal bracing members should directly connect with the gable or separating walls. The lower ends of diagonal rafter bracing members should abut the end walls as closely as possible to the intersection between the wall and the wallplate.

BRACING
continued.

Where a roof has a long span, bracing additional to the standard longitudinal and rafter diagonal bracing becomes necessary. The illustrations on this page and the following pages highlight such additional bracing. As for standard bracing, bracing timber should be 100x25mm nominal size, free of defects and twice nailed with 75mm galvanised nails of 2.65mm diameter at each point of contact with the truss.

The following table sets out the conditions under which bracing additional to the minimum standard is required.

		Roof pitch in degrees					
		22.5	25	27.5	30	32.5	35
Brace internal tie where span exceeds		10.2 m	9.5 m	8.7 m	8.0 m	7.5 m	7.0 m
Brace internal strut where span exceeds		—	—	—	—	10.7 m	10.0 m

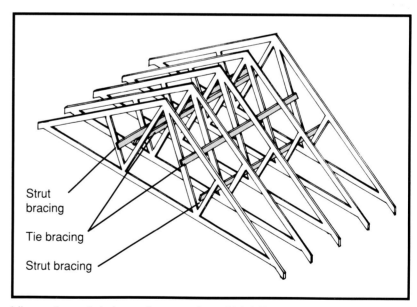

Strut bracing

Tie bracing

Strut bracing

View of pre-fabricated trussed roof highlighting additional bracing to internal truss members.
(See also additional note on page 138)

BRACING
continued.

In addition, where it is necessary to brace ties or struts as illustrated on the previous page it is also required that additional bracing timbers (chevron bracing) are fixed as illustrated in the sketches opposite, repeating without interruption along the length of the roof.

BRACING SHOULD NOT CROSS A PARTY WALL

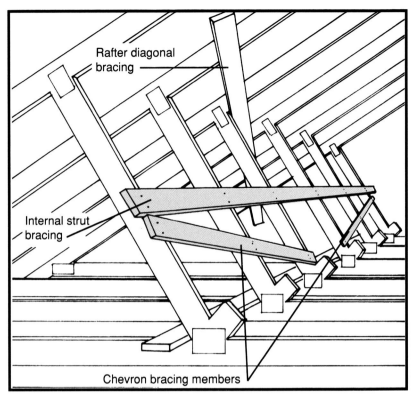

Rafter diagonal bracing

Internal strut bracing

Chevron bracing members

Chevron bracing to internal strut.

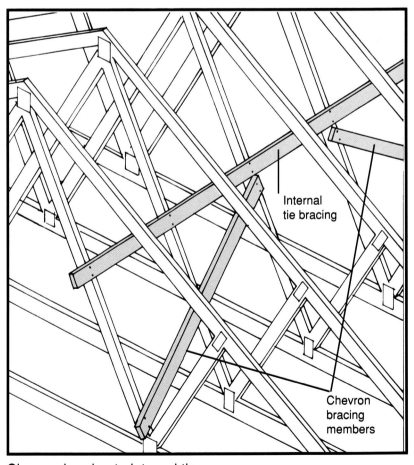

Internal tie bracing

Chevron bracing members

Chevron bracing to internal tie.
Note: In the interest of clarity some bracing members have been omitted from this sketch.

STRAPPING

Strapping of trusses to gable walls. To help stabilise gable walls, such walls should be tied to the roof timbers as illustrated.

Connections should be made to the rafter with 30x5 thick galvanized steel straps fixed to at least 2 trusses and a nogging with 3.35x50 long wire nails.
On gable walls they should be spaced at not more than 2m centres at rafter and ceiling tie level.

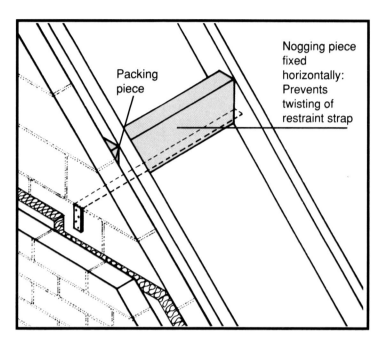

Straps at rafter level: The turn down of the strap must be tight against the block.

Restraint straps: to gable walls at rafter and ceiling level at not more than 2m centres.
Note: Straps are not required at party wall.

NOTE: THE STRAPS MUST BE BUILT IN BY THE BLOCK-LAYER AS THE WALL RISES.

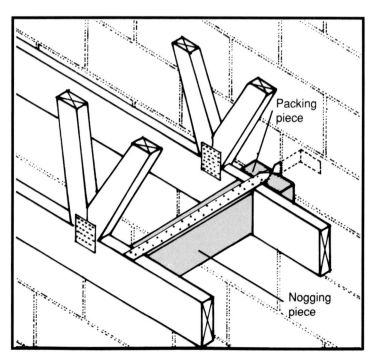

Straps at ceiling level and gable wall. The requirement for strapping at ceiling level depends on height of apex and wall thickness as specified in Technical Guidance Document A of the Building Regulations. In general it is required in all standard pitched roof houses.

PARTY WALLS

At party walls between houses the construction should take account of deflection and the need for fire stopping.

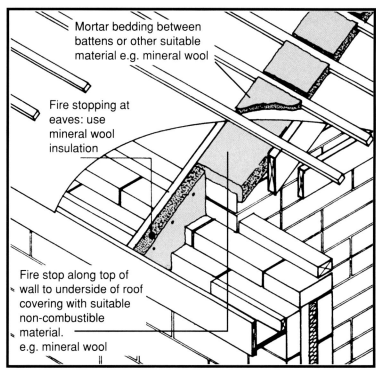

Mortar bedding between battens or other suitable material e.g. mineral wool

Fire stopping at eaves: use mineral wool insulation

Fire stop along top of wall to underside of roof covering with suitable non-combustible material.
e.g. mineral wool

Eaves level

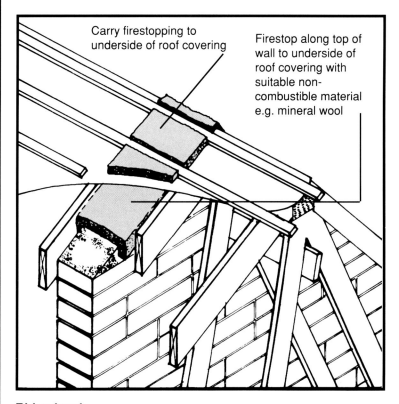

Carry firestopping to underside of roof covering

Firestop along top of wall to underside of roof covering with suitable non-combustible material e.g. mineral wool

Ridge level:
Top of rafter should be 25mm approx. above top of blockwork to allow for deflection of roof timbers when fully loaded with tiles or slates. This 25mm gap should be packed with mineral wool to seal the gap against fire spread.

PARTY WALLS
continued.

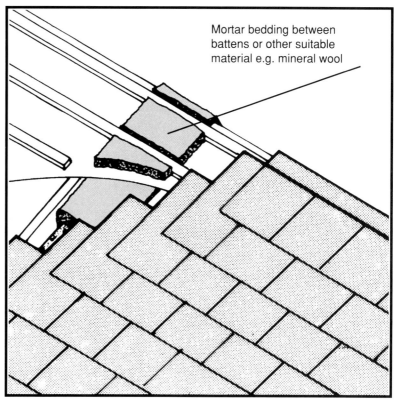

Mortar bedding between battens or other suitable material e.g. mineral wool

The completed party wall with felt, battens, and slates or tiles carried over the fire stopping.

HATCH AND CHIMNEY OPENINGS

Where it is not possible to accommodate hatches and chimneys in the standard spacing between trusses the following increased truss spacing guidelines illustrated here and recommended by the truss manufacturers should be adhered to. This will ensure that no individual trussed rafter is subject to a load greater than that applied, were it at standard spacing.

Definitions:
S: Standard truss spacing (normally 600).
C: Increased truss spacing.
B: Reduced truss spacing.

Condition 1: Illustrates a spacing adjustment which involves increasing the spacing between two trusses by up to approx. 10% (i.e. up to approx. 660mm). This increase causes no significant overstressing of the tiling battens or ceiling material.
See example 1 on page 144.

Condition 2: Illustrates spacing adjustments for openings greater than 10% standard spacing and up to twice standard spacing (i.e. between 660mm and 1200mm). Tiling battens and ceiling material should be given extra support provided by infill rafter(s) and ceiling joist(s). Support of the infill timber is provided in line with each truss joint by a purlin, binder or ridge board and by trimmers at the actual opening, as illustrated on page 143. See example 2 or 3 on pages 144 and 145.

NEVER CUT A PREFABRICATED TRUSS

Condition 1

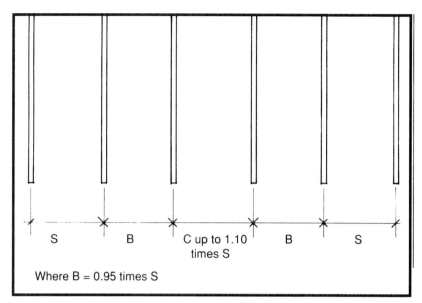

Where B = 0.95 times S

Up to 660mm approx. (for trusses at 600mm centres)

Condition 2

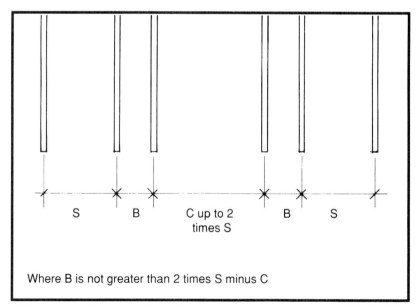

Where B is not greater than 2 times S minus C

Between 660mm and 1200mm
(for trusses at 600mm centres)

e.g. If C = 1000
B = 200

HATCH AND CHIMNEY OPENINGS
continued.

Condition 3: Illustrates spacing adjustments for openings greater than twice standard truss spacing and up to three times standard truss spacing (i.e. between 1200mm and 1800mm) providing two trimming trusses at each side of the opening are used and are nominally fixed together with nails at 300mm centres along all members. Tiling battens and ceiling material should be given extra support provided by infill rafters and ceiling joists. Support of the infill timber is provided in line with each truss joint by a purlin, binder or ridge board and by trimmers at the actual opening, as illustrated on page 146.

Note: Where circumstances dictate a spacing greater than three times standard truss spacing (i.e. greater than 1800mm) then design by an engineer is required. The engineer appointed should be qualified by examination, be in private practice and possess professional indemnity insurance.

Condition 3

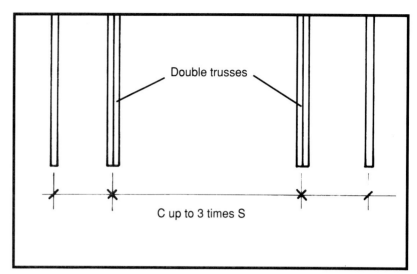

Between 1200mm and 1800mm (for trusses at 600mm centres)

Note: Where truss spacing exceeds 1200mm double trusses nominally fixed together with nails at 300mm centres along all members must be used on either side of the opening.

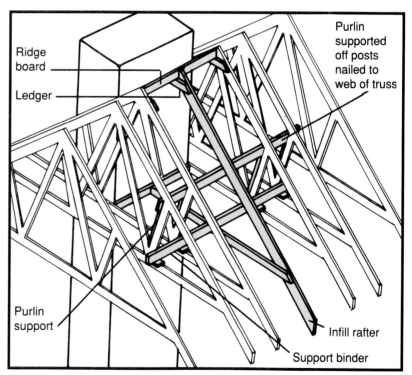

Support of the infill timber is provided in line with each truss joint by a purlin, binder or ridge board and by trimmers at the actual opening.

Note: Care should be taken when nailing the purlin and purlin support to the trussed rafter strut to avoid damage.

HATCH AND CHIMNEY OPENINGS
continued.

The following pages illustrate how the conditions for increased truss spacings as outlined on the previous pages should be applied.

Note: The span of any (44 x 35mm) tiling or slating batten should not exceed 670mm. To achieve this for the situation shown opposite, nail 44 x 35mm battens to the last truss on either side of the party wall as illustrated on the following page. This will reduce the batten span to 670mm.

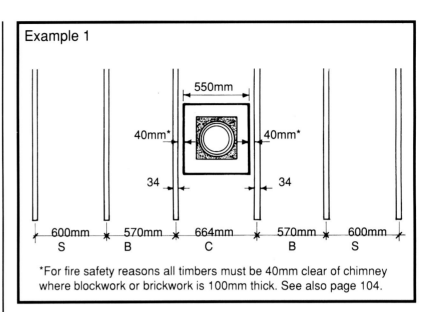

Example 1

550mm

40mm*   40mm*

34   34

| 600mm | 570mm | 664mm | 570mm | 600mm |
| S | B | C | B | S |

*For fire safety reasons all timbers must be 40mm clear of chimney where blockwork or brickwork is 100mm thick. See also page 104.

Maximum truss spacing dimensions for standard block chimney with a single flue serving one house, as derived from condition 1 on the previous page. Where increased spacing (C) can be up to 1.10 times standard spacing (S) and reduced truss spacing (B) should not be greater than 0.95 times standard spacing (S).

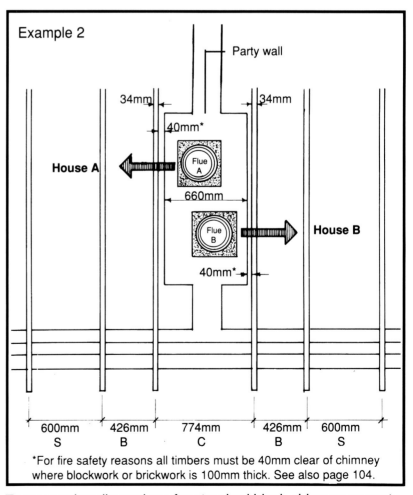

Example 2

Party wall

34mm   34mm

40mm*

House A

Flue A

660mm

Flue B   House B

40mm*

| 600mm | 426mm | 774mm | 426mm | 600mm |
| S | B | C | B | S |

*For fire safety reasons all timbers must be 40mm clear of chimney where blockwork or brickwork is 100mm thick. See also page 104.

Truss spacing dimensions for standard block chimney on party wall (with two flues) as derived from condition 2 on the previous page.

HATCH AND CHIMNEY OPENINGS
continued.

Note: Where infill rafters and ceiling joists are required support for these members should be provided in accordance with the guidance on page 146.

Example 2
continued.

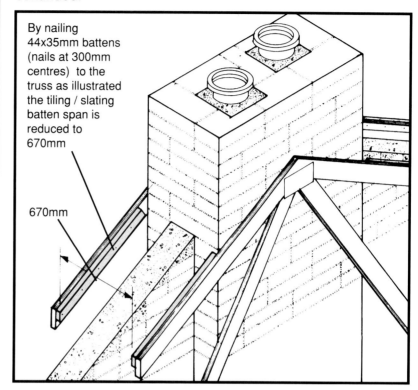

By nailing 44x35mm battens (nails at 300mm centres) to the truss as illustrated the tiling / slating batten span is reduced to 670mm

670mm

Detail at party wall to avoid excessive spans of tiling / slating battens.

Example 3

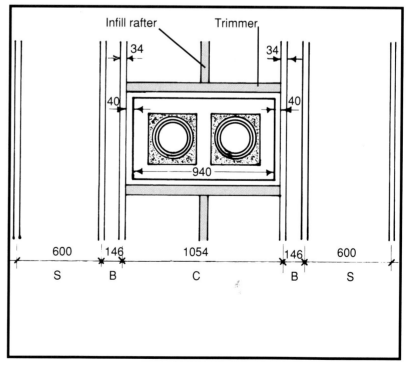

Truss spacing dimensions for a standard block chimney with two flues, both within same house. i.e. not on a party wall as derived from condition 2 on page 142.

TRIMMING AROUND CHIMNEYS
continued.

Where truss spacings are increased to accommodate chimneys in some cases trimming may be unavoidable.

Where this occurs then increased truss spacing should be in accordance with the preceding pages. The sketches on the following pages illustrate the correct procedure for providing support to the infill timbers which are required to provide support to the tiling battens and ceiling materials.

Note:
All timbers must be 40mm clear of chimney where blockwork or brickwork is 100mm thick for fire safety reasons. See also page 104.

Infill timbers sized in accordance with SR11 and at least 25mm deeper than trussed rafter.

DO NOT NAIL TIMBER DIRECTLY TO CHIMNEY

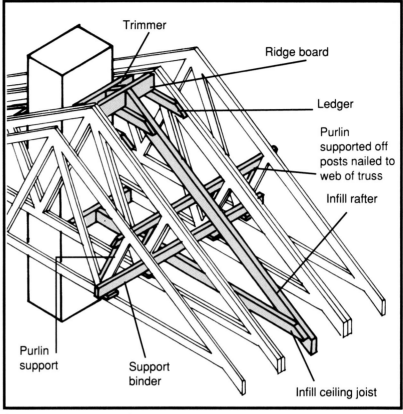

Support to loose timbers

Care should be taken when nailing the purlin and purlin support to the trussed rafter strut to avoid damage.

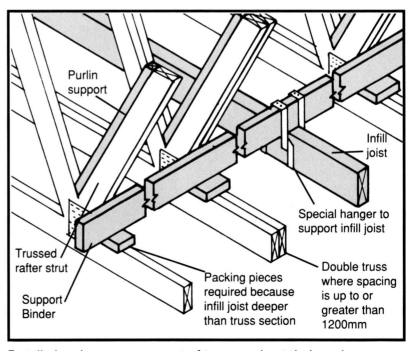

Detail showing arrangement of truss and cut timbers in a typical trimming arrangement around stack.
Note that the loads on the infill timbers are transferred to the double-trusses either side and also to the next adjoining truss.

**TRIMMING AROUND
CHIMNEYS**
continued.

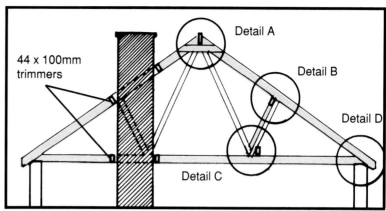

Typical section through chimney trimming
Note: This sketch is not illustrating a cut truss, it is a section
through the infill timbers between trusses.

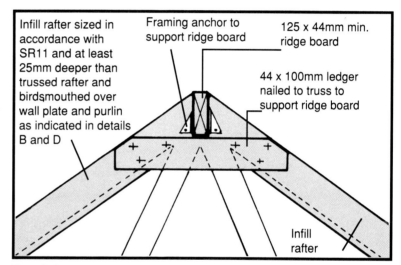

Detail A

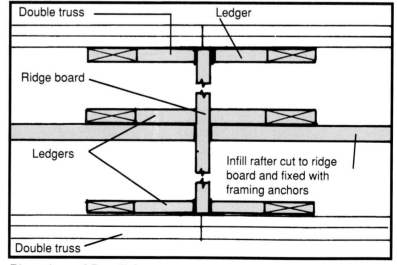

Plan view of Detail A

TRIMMING AROUND CHIMNEYS
continued.

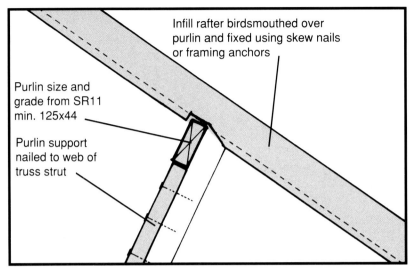

Infill rafter birdsmouthed over purlin and fixed using skew nails or framing anchors

Purlin size and grade from SR11 min. 125x44

Purlin support nailed to web of truss strut

Detail B

As an alternative to using loose timbers as trimming a specially designed and fabricated 'stubbed' truss can be used as illustrated below.

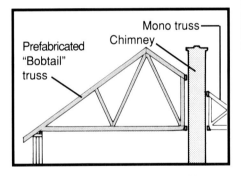

Prefabricated "Bobtail" truss

Mono truss

Chimney

The bobtail truss is specifically designed and fabricated for such situations and is not adapted from an ordinary truss by site cutting. The bobtail truss should be erected in accordance with the manufacturer's instructions.

All timbers to be kept 40mm clear of chimney.

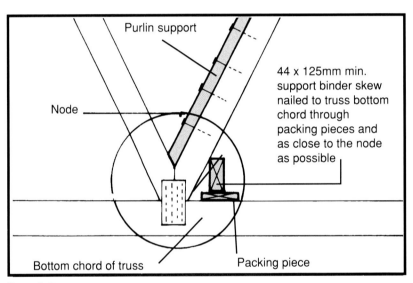

Purlin support

Node

44 x 125mm min. support binder skew nailed to truss bottom chord through packing pieces and as close to the node as possible

Bottom chord of truss

Packing piece

Detail C
Note: Support binder size and grade designed to suit span.

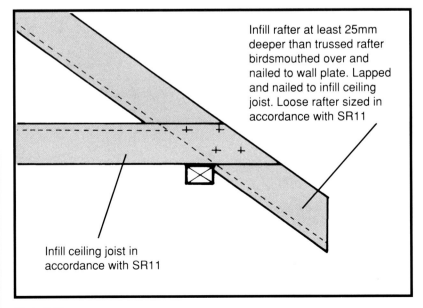

Infill rafter at least 25mm deeper than trussed rafter birdsmouthed over and nailed to wall plate. Lapped and nailed to infill ceiling joist. Loose rafter sized in accordance with SR11

Infill ceiling joist in accordance with SR11

Detail D

WATER CISTERNS

It is essential that the weight of the water cistern be spread over a number of trusses.
Failure to do this will lead to local deflection and cracking of the ceiling below the cistern.
Dimensions for the support members can be taken from the table on this page.

Water storage capacity:

Minimum 212 litres for a three bedroomed house; minimum 340 litres for four bedroomed or larger houses. The requirements of individual Local Authorities may vary and should be checked.

Cistern cover:

The cistern should be covered, but not air–tight.

Note:
For cisterns with a nominal capacity greater than 270 litres, special design is required by I.S. 193: 1986.

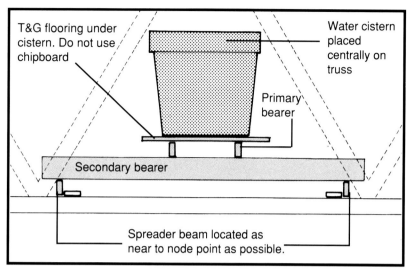

Water cistern support.

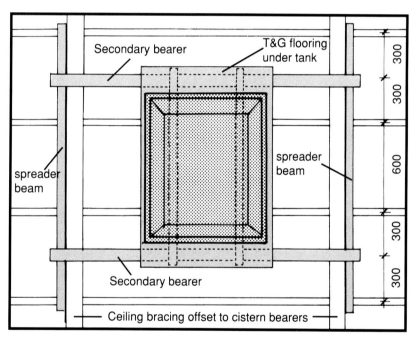

Cisterns with a nominal capacity of up to 270 litres should be spread over **four** trusses. Where more than one cistern is required, no truss should carry the load from more than one cistern.

Limit of span trussed rafter	Primary bearers	Secondary bearers	Spreader beams
m	mm	mm	mm
8.0	35 x 100	44 x 175	44 x 100
11.0	35 x 100	75 x 150	44 x 100

Minimum size of cistern support members, for capacity up to 270 litres nominal

INSULATION AROUND TANKS

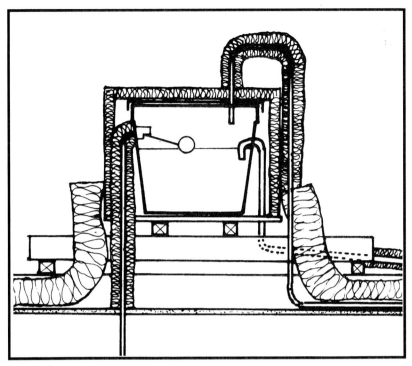

Insulation around water tank and
associated pipe work

Note: Insulation omitted under tank, to
expose water to heat from within house
and reduce risk of freezing.

Trap doors:
The standard spacing between
prefabricated trusses will
accommodate the trap door.
Under no circumstances should
the trussed rafter be cut to make
space for a larger opening.

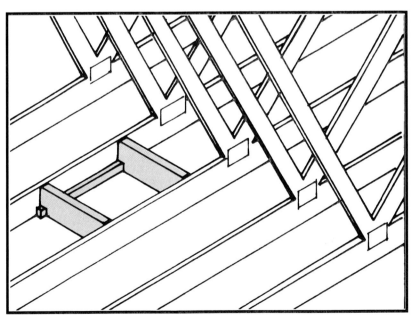

Trap door accommodated in standard
spacing between trusses. Framing
members notched over battens and fixed.

GABLE LADDERS

Gable ladders are used where a roof overhang is required at a gable end, and they should be securely nailed directly to the last truss.

The gable ladder should be evenly supported by the gable blockwork. Barge boards and soffits can be nailed directly to the gable ladder. For gable overhangs in excess of 300mm from the outside face of the wall, special design will be required.

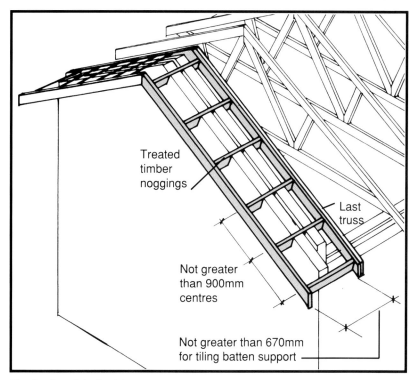

Treated timber noggings

Last truss

Not greater than 900mm centres

Not greater than 670mm for tiling batten support

Typical gable ladder.

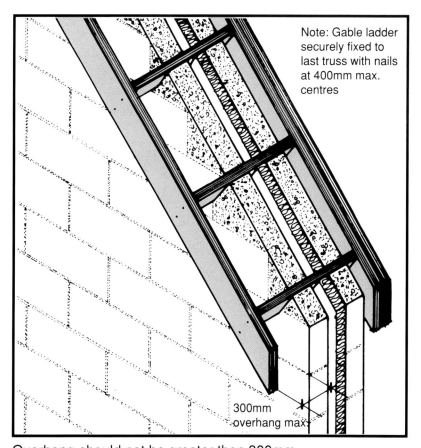

Note: Gable ladder securely fixed to last truss with nails at 400mm max. centres

300mm overhang max

Overhang should not be greater than 300mm

FORMING HIPPED ENDS

Forming hips in a roof with prefabricated trusses can be done in two ways.

1   Using cut timbers to form the hipped section of the roof and/or

2   Using specially designed and manufactured mono trusses to suit the roof.

Note:
If using a cut hipped end where loading is transferred to a load bearing partition wall, then this wall or partition must have a separate foundation because it is load bearing.

Sizes of ceiling joists and rafters from SR11.

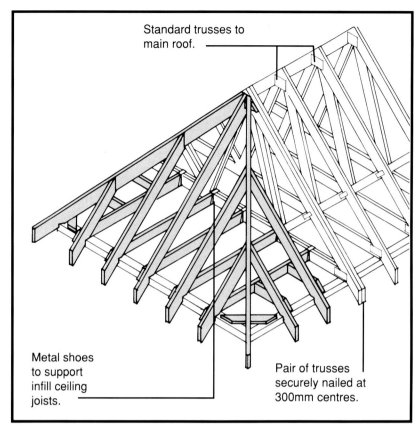

Standard trusses to main roof.

Metal shoes to support infill ceiling joists.

Pair of trusses securely nailed at 300mm centres.

1

View of a hipped roof with prefabricated trusses and cut timber hips.
Note: This method should only be used where the length of the gable wall does not exceed 5m. Where this dimension exceeds 5m the method of forming the hipped ends illustrated on page 153 should be used.

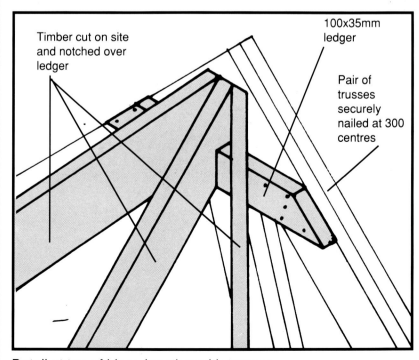

Timber cut on site and notched over ledger

100x35mm ledger

Pair of trusses securely nailed at 300 centres

Detail at top of hips: Junction with trusses.

FORMING HIPPED ENDS
continued.

Option 2:
Using specially designed girder truss and mono trusses together with cut infill timber to form the hip.

Sizes of ceiling joists and rafters from SR11.

If using this method to form the hipped end the following should be taken into consideration.

◆ Ensure that the girder truss* contains the correct number of flat top plys (usually 2 or 3 nailed together) in accordance with the manufacturer's instructions.

◆ Brace the top chord of the flat top trusses to the girder truss.

◆ Use truss shoes to support the mono trusses at the girder truss.

◆ Birdsmouth the hip rafter over the girder truss to ensure support.

* Girder truss: A truss comprising two or more individual trusses (plys) fixed together and designed to carry exceptional loads such as those imparted by other trusses.

2

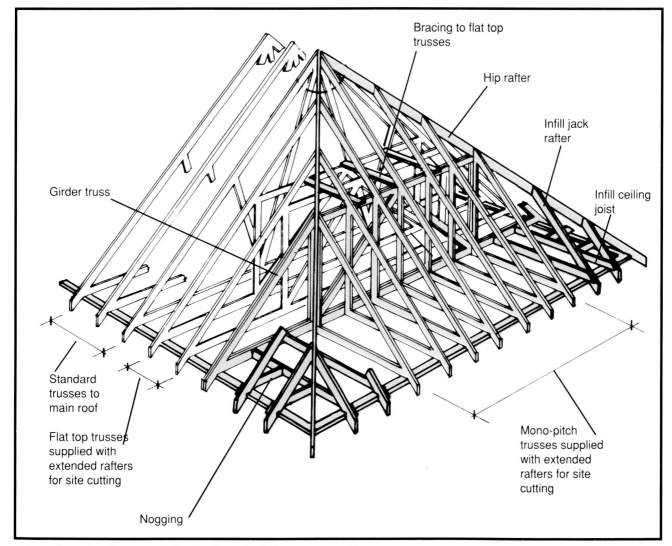

Hipped roof using specially designed girder truss and mono trusses together with cut infill timber to form the hip.

FORMING HIPPED ENDS
continued.

A large proportion of the load of a hipped roof is transmitted to the walls at the corner. To avoid "kicking" of the roof at the feet of hips, care should be taken with the fixing and joining of the wallplate. For further details of good practice in this area see pages 159 to 161.

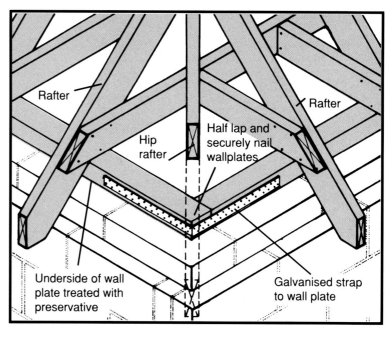

Rafter

Rafter

Hip rafter

Half lap and securely nail wallplates

Underside of wall plate treated with preservative

Galvanised strap to wall plate

Typical detail at foot of hip. (Showing cut-away hip rafter). In heavily loaded hips an angle tie and dragon tie should be used as an alternative. See page 161.

HANDLING OF TRUSSES

Trusses are designed to be loaded in a particular way when erected. They must be handled carefully so as not to damage the fixing of the truss plates. They must be stored carefully (see page 224) and when being erected they must be put into place without straining.

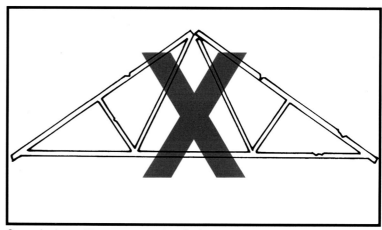

Care in handling trusses is vital:
REJECT ANY DAMAGED OR DISTORTED TRUSSES

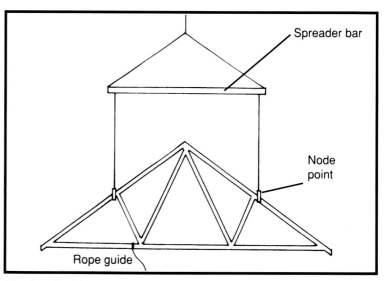

Spreader bar

Node point

Rope guide

Mechanical handling technique

INTRODUCTION

The essence of good traditional cut roof construction can be summarised as:

◆ Properly sized timber - use SR11 to select timber sizes.

◆ Full triangulation of the roof - if the joists and rafters do not meet at wall plate level, there is a risk of roof spread and an engineer's design is required to avoid this risk. **Inadequate triangulation is a significant source of problems in cut roofs.** The engineer appointed must be qualified by examination, be in private practice and possess professional indemnity insurance.

◆ Proper transfer of purlin loads either to load bearing walls or specially designed joists.

Proper cut roof construction. The sketches on these pages show the parts of a well designed and constructed cut roof. Note the triangulation at eaves level, and the purlin strutted on to load bearing wall.

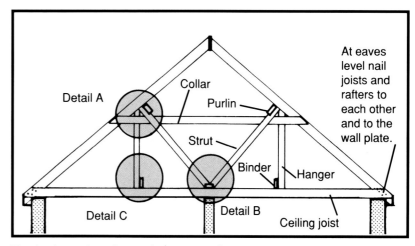

Typical section through a cut roof.

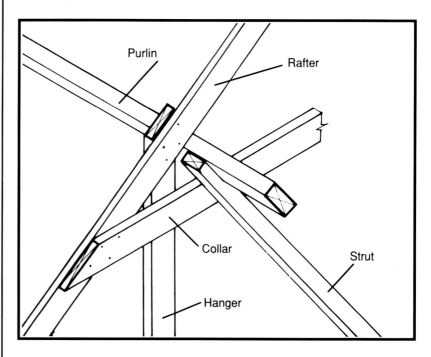

Detail A: Arrangement at junction of rafter, purlin hanger and collar.

**PROPER CUT ROOF
CONSTRUCTION**
continued.

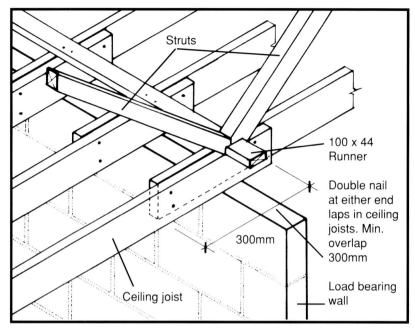

Struts

100 x 44
Runner

Double nail
at either end
laps in ceiling
joists. Min.
overlap
300mm

300mm

Load bearing
wall

Ceiling joist

Detail B: Arrangement of struts at ceiling level.

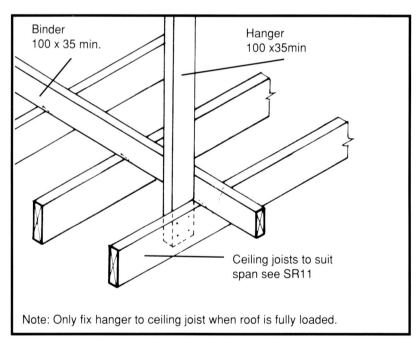

Binder
100 x 35 min.

Hanger
100 x35min

Ceiling joists to suit
span see SR11

Note: Only fix hanger to ceiling joist when roof is fully loaded.

Detail C: Hanger and joist arrangement.

Note: Traditional cut roof construction
assumes that the binder/hanger supports
the ceiling joist. The binder is securely fixed
to each ceiling joist and the hangers which
are provided every third or fourth rafter are
nailed to the ceiling joists after the roof is
loaded.

PURLINS

Purlins are horizontal members which give intermediate support to the rafters. The purlins are in turn supported by struts which must bear onto a load bearing partition or specially designed ceiling joist.

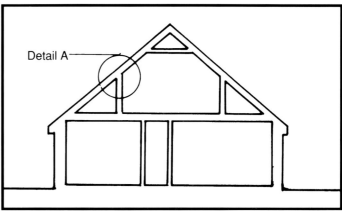

Key sketch – location of purlin.

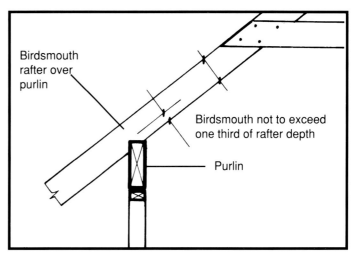

Detail A
Rafters to purlin: A birdsmouth joint should be used if the purlin is to be fixed vertically

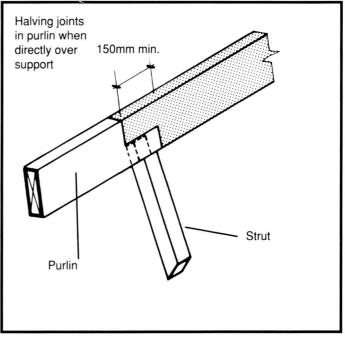

Purlin connections: Support should be provided directly under joint.

PURLINS
continued.

It is essential to prop purlins at regular intervals.

For example, a 175 x 75mm purlin of Strength Class A with a rafter span of 1.75m and rafter pitch of 30° would normally require to be propped every 2.24m and a 225 x 75mm purlin of Strength Class C with a rafter span of 1.5mm would normally require to be propped every 3.37m.

Note: The rafter span is measured on plan.

Use SR11 to properly size purlins (see pages 185 to 187).
NEVER PROP A PURLIN OFF A CEILING JOIST UNLESS THE CEILING JOIST IS DESIGNED TO CARRY THE LOAD.

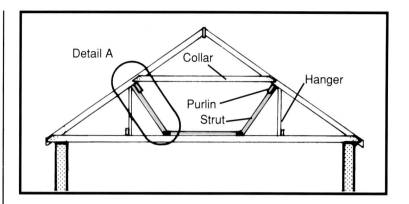

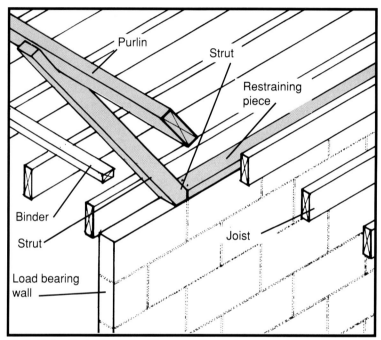

Detail A: Detail of strutting purlin onto a load bearing wall
Note: For clarity the hangers have been omitted from this sketch.

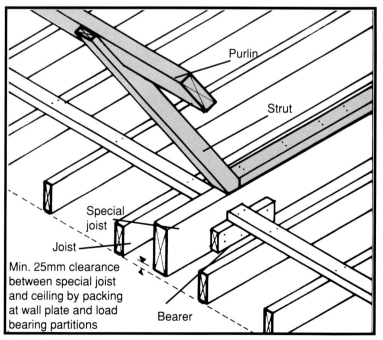

Alternative detail A: Where a purlin cannot be propped onto a wall it may be propped on a specially designed timber, **not** an ordinary ceiling joist, or an RSJ or UB, see page 163.

HIPPED ENDS

See comment on pages 152 and 153 with regard to the construction of hips.

Hip rafter and purlin connection:
It is vital that the hipped rafter in this location is notched over and securely nailed to the purlins. The purlins must be mitred and supported at or close to their intersection and the struts themselves supported.

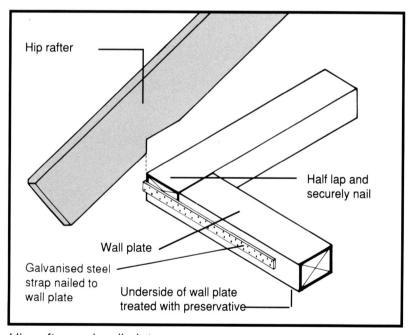

Hip rafter

Half lap and securely nail

Wall plate

Galvanised steel strap nailed to wall plate

Underside of wall plate treated with preservative

Hip rafter and wall plate

HIPPED ENDS
continued.

To reduce outward thrust at the corner under the hip, it is necessary at the intersection of the wall plates to half lap and securely nail them and then:

1. Reinforce the corner with a galvanised steel strap as illustrated in Fig. 1.
   or
2. Reinforce the corner with an angle tie securely nailed to the wall plates as illustrated in Fig. 2.

   Additionally in case of heavily loaded hips securely fix a galvanised steel dragon tie to both angle tie and hip rafter as illustrated on page 161.

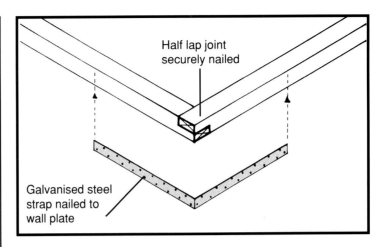

1

Use of galvanised steel strip to reinforce corner. This method is strongly recommended.

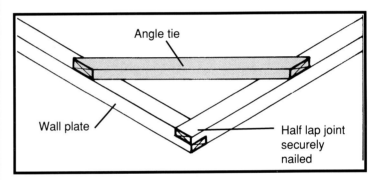

2

Angle tie nailed to wall plates and used in conjunction with a galvanised steel dragon tie to reinforce the corner in the case of heavily loaded hips. (Note: In the interest of clarity the galvanised steel dragon tie has been omitted from this sketch. (See page 161)

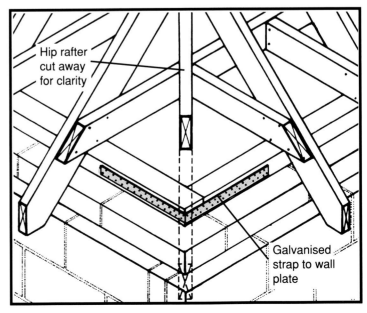

'Cutaway' view of typical detail at foot of hip.

HIPPED ENDS
continued.

Hips: See commentary on pages 152 and 153 with regard to the construction of hips.

Angle tie at corner – note wall plate tying down straps.

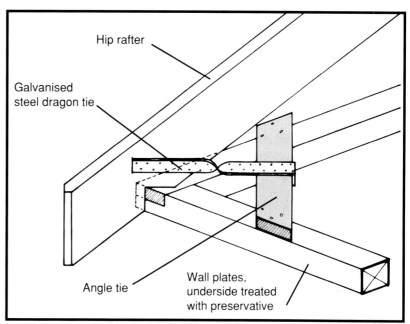

Hip rafter

Galvanised steel dragon tie

Angle tie

Wall plates, underside treated with preservative

Use of a galvanised steel dragon tie securely fixed to the hip rafter and the angle tie, in the case of heavily loaded hips.

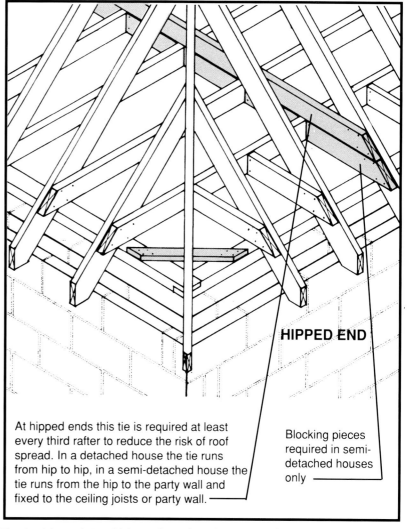

HIPPED END

At hipped ends this tie is required at least every third rafter to reduce the risk of roof spread. In a detached house the tie runs from hip to hip, in a semi-detached house the tie runs from the hip to the party wall and fixed to the ceiling joists or party wall.

Blocking pieces required in semi-detached houses only

Typical junction of hipped roof and corner of wall.
As stated earlier the corner under the hip is subject to outward thrust from the roof load. Secure fixing and tying of the wall plate is necessary to satisfactorily resist this thrust.

VALLEY CONSTRUCTION

When constructing valley roofs, particular care should be exercised in the weathering and structural detailing. Where a single valley rafter supports the jack rafters, significant bending moments in the valley rafter and horizontal thrust at eaves level can arise. These forces and moments can be catered for, along with valley/rafter connections. Where the roof rafters are supported by intermediate purlins, the valley rafter often bears on the junction of the two purlins. Consideration must be given to the support of the purlins at this point. Where any doubt exists about the structure of the valley an engineer should be engaged, the engineer appointed should be qualified by examination, be in private practice and possess professional indemnity insurance.

For guidance on the weathering of valleys see pages 197 and 198.

WATER CISTERN SUPPORT IN CUT TIMBER ROOFS

As with prefabricated trussed roofs the weight of the water cistern in cut timber roofs should be adequately supported.

Where possible the cisterns should be located directly above and bear on a load bearing partition. Where this is not possible the joists carrying cisterns must be specifically designed to carry the additional load.

INTRODUCTION

The incorporation of accommodation within the roof space is a common feature in house design. A roof incorporating such accommodation is called a dormer roof and the term dormer bungalow similarly applies to a bungalow with rooms in the roof space.

The construction of a dormer roof requires that, in addition to the matters relevant to the conventional roof as outlined on the preceding pages, a number of other items require attention.

Structure.

In dormer roofs the need to provide clear space of adequate width and height to accommodate rooms means that the structural form of the roof differs significantly from a conventional roof space. The design of the roof structure must take this into account, and ensure that sizing and disposition of members is adequate, and must take into account the additional loads imposed by the use of the roof space as domestic accommodation, particularly on joists. In some cases a satisfactory structure can be achieved by timber construction throughout. In other cases it may involve the incorporation of structural steel members such as RSJ's or Universal Beams as part of the structural design of the roof.

Partially constructed dormer roof, incorporating steel sections as part of roof structure.

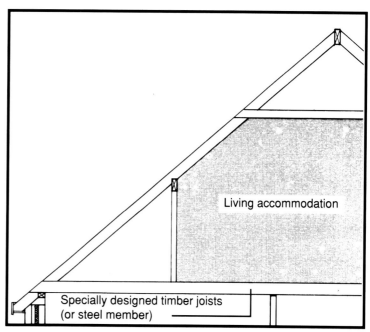

Dormer roof structure incorporating timber structural members. In this case the joists must be designed to take extra load from the roof or else an RSJ or U.B. should be provided under purlin support.

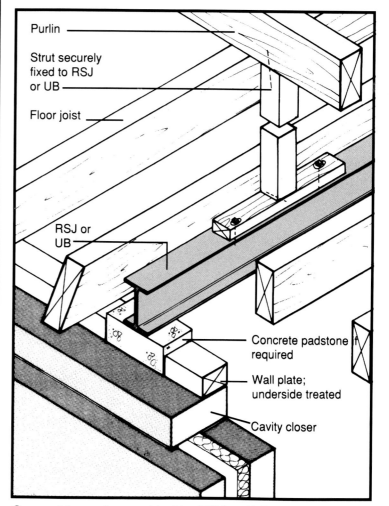

Support to purlin provided by RSJ or U.B.

STRUCTURE
continued.

As a general rule the overall shape of a dormer roof should form a triangle, that is, the floor joists should be tied to the feet of the rafters to form a rigid frame. In some cases this may not occur – for example, where wall plate level is raised above floor level of the dormer accommodation to achieve a satisfactory ceiling height. In such cases, the absence of the usual triangulation must be compensated for by suitable alternative means in the design of the roof structure.

In all cases the structural design of a dormer roof should be carried out by an engineer. The engineer appointed should be qualified by examination, be in private practice and possess professional indemnity insurance

As an alternative to an engineer designed cut roof, purpose–designed prefabricated dormer trusses can be used to form the roof structure.

Daylighting in dormer roofs. The provision of day light to rooms in dormer roofs is commonly achieved by the incorporation of dormer windows or rooflights. While there are no specific daylighting provisions in Building Regulations for window/rooflight sizes, it is recommended that the area of such openings be at least 10% of the floor area of the room served, to ensure a reasonable level of daylighting.

Ventilation is required to ensure that dormer roofs are adequately ventilated to prevent condensation. A vapour barrier is also required on the warm side of the insulation, (See page 169)

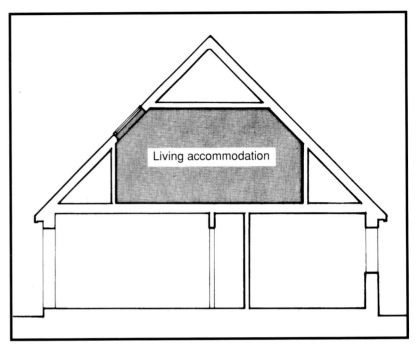

Living accommodation

Dormer roof incorporating conventional triangulation at eaves

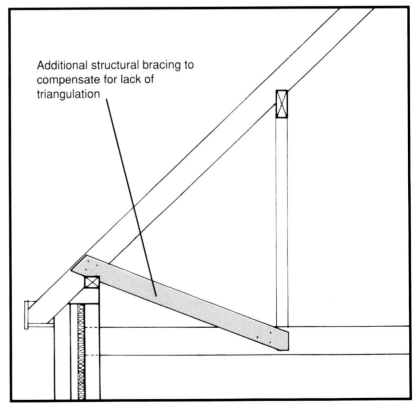

Additional structural bracing to compensate for lack of triangulation

Dormer roof with wall plate level above joist level – structural bracing of the type illustrated or suitable alternative must be provided to compensate for absence of triangulation. Engineer's design required in these situations. The engineer appointed should be qualified by examination be in private practice and posses professional indemnity insurance.

DORMER WINDOWS

Ensure dormer windows of the type illustrated opposite are fully framed and jointed before roof finishes to dormer and main roof are laid.

Trimming members around dormers should be of adequate size to take the extra load from the cut main roof members and dormer framing and cladding as detailed in the design.

The dormer window structure should be framed so that they are independent of the window frame, using a suitable lintel over the opening. Window frames are not intended to carry roof loads.

Weathering details
Weathering at junction of pitched roof and flat roof to dormer window. Where a flat roof adjoins a pitched roof, or where valleys or gutters occur, the water proof membrane should be carried up under the tiling to a height of 150mm above the flat roof, valley or gutter and overlapped by the roofing underlay.

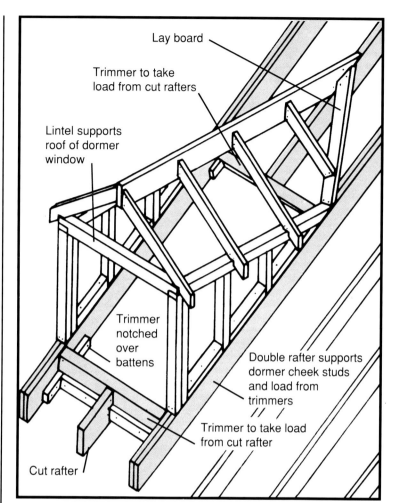

Dormer window construction.

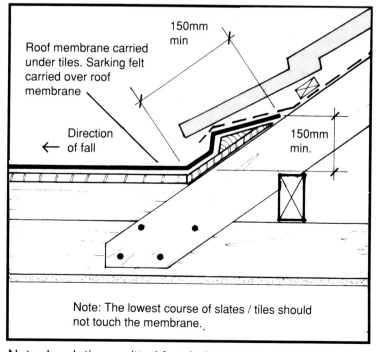

Note: Insulation omitted for clarity.

DORMER WINDOWS
continued.

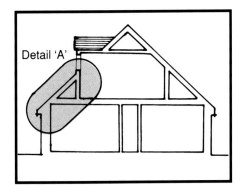

Key sketch

Note:

Every effort should be made at design stage to ensure rafters and floor joists run in the same direction so that adequate triangulation can be achieved. If due to layout of joists this is not possible, then engineer's advice must be obtained to ensure that there is adequate tying back of rafters as indicated in the sketch. The horizontal thrust must be catered for to stop roof spread.

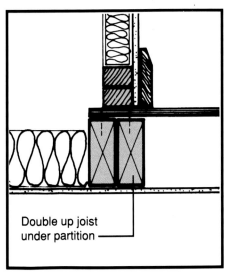

Double up joist under partition

Floor boards act as bracing.

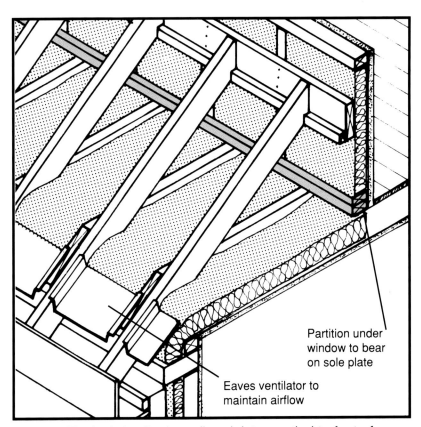

Partition under window to bear on sole plate

Eaves ventilator to maintain airflow

Detail A: Typical detail where floor joists are tied to feet of rafters. The floor joists must be designed to take the load.

Alternative detail 'A'

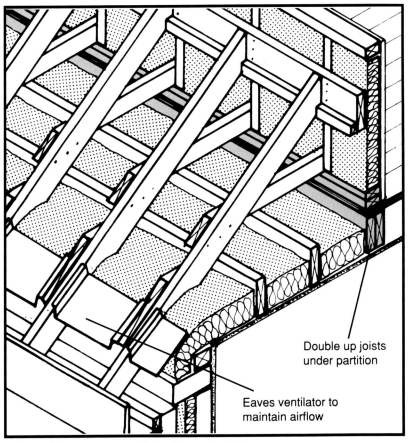

Double up joists under partition

Eaves ventilator to maintain airflow

Where joists run parallel with wall plate.
Partition to bear on pair of joists.

INTRODUCTION

Ventilation of roof spaces is necessary to remove water vapour and prevent harmful condensation. The need for ventilation is a consequence of higher standards of insulation, and an increase in activities generating water vapour within buildings.

Roof ventilation requirements

The sketches on pages 167 to 169 illustrate the dimensional requirements for continuous ventilation openings required for various roof types, as required by Technical Guidance Document F of the Building Regulations. These requirements should be complied with to avoid long–term condensation and associated problems.

The continuous openings or their equivalent* should have an area on opposite sides at least equal to a continuous ventilation strip running the full length of the eaves. The width of the strip depends on the roof type.

*The sketches on pages 167 to 169 illustrate the roof void being ventilated through the soffit. If, however, it is not possible to do this, alternative means of ventilation may be adopted, e.g. the use of proprietary vent slates or vent tiles, providing the number of vent slates/tiles used provides the same equivalent area of ventilation as would ventilating through the soffit for that particular roof type. See page 171 for additional information and calculation methods for vent slates/tiles.

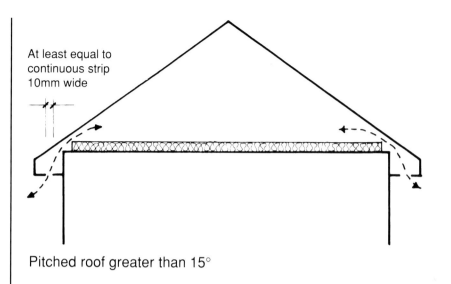

Pitched roof greater than 15°

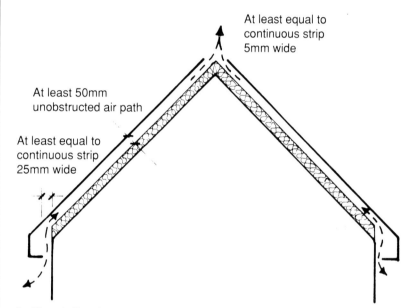

Ceiling following pitch of roof.

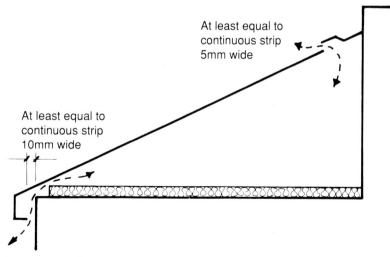

Lean to roof

Note: Dimensions for ventilation openings given in all the illustrations are for continuous openings. Alternatively regularly spaced openings giving the same aggregate area may be used.

ROOF VENTILATION REQUIREMENTS
continued.

Means of ventilation

Continuous eaves ventilation using proprietary soffit vent incorporating fly screen to prevent birds, insects etc. entering the roof void.

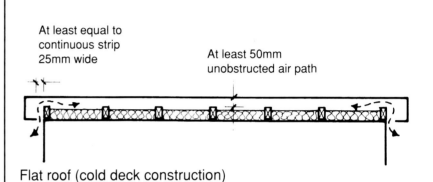

At least equal to continuous strip 25mm wide

At least 50mm unobstructed air path

Flat roof (cold deck construction)

At least equal to continuous strip 25mm wide

At least 50mm unobstructed air path

Flat roof (cold deck construction)

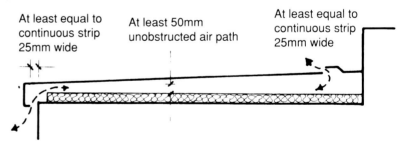

At least equal to continuous strip 25mm wide

At least 50mm unobstructed air path

At least equal to continuous strip 25mm wide

Flat roof abutment with wall (cold deck construction)

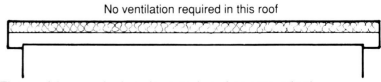

No ventilation required in this roof

Flat roof (warm deck or inverted roof construction)

Note: Dimensions for ventilation openings given in all the illustrations are for continuous openings alternatively regularly spaced openings giving the same aggregate area may be used.

ROOF VENTILATION REQUIREMENTS
continued.

Vapour control layers:

Vapour control layers can reduce the amount of moisture reaching the roof void but they cannot be relied on as an alternative to ventilation. In the case of dormer roofs where the ventilation is along the line of the rafters, it is necessary to install a vapour control layer on the warm side of the insulation, e.g., 500 gauge polythene with sealed laps or its equivalent.

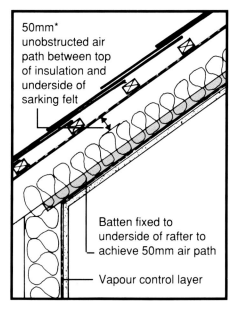

50mm* unobstructed air path between top of insulation and underside of sarking felt

Batten fixed to underside of rafter to achieve 50mm air path

Vapour control layer

*To ensure the provision of the 50mm unobstructed air path as illustrated on the sketches opposite, it may be necessary because of the depth of insulation required and the depth of the rafter to batten out the sloping portion of the roof. Alternatively, deeper rafters than actually required may be used.

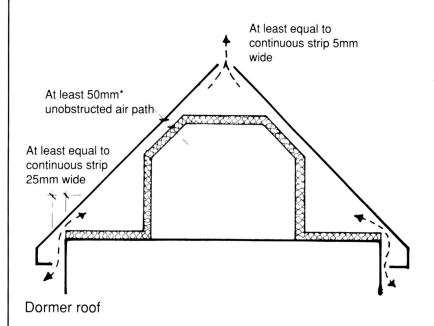

At least equal to continuous strip 5mm wide

At least 50mm* unobstructed air path

At least equal to continuous strip 25mm wide

Dormer roof

Note: If the roof above a dormer window is of pitched roof construction provide at least 10mm continuous strip along eaves on both sides.

At least equal to continuous strip 5mm wide

At least 50mm* unobstructed air path

At least equal to continuous strip 25mm wide

At least equal to continuous strip 5mm wide

At least equal to continuous strip 25mm wide

Dormer roof with flat roof dormer window.

Note: Dimensions for ventilation openings given in all the illustrations are for continuous openings. Alternatively regularly spaced openings giving the same aggregate area may be used.

MEANS OF VENTILATION

Eaves ventilation.
To ensure an unobstructed flow of air over the insulation at eaves level, the use of proprietary eaves ventilators is recommended.

Typical eaves ventilators

Fixing of eaves ventilators.

Tile/slate ventilators.
Where it is not possible to ventilate the roof void through the soffit or at a lean to roof abutment, proprietary tile/slate ventilators may be used. Page 171 gives guidance on how to calculate the number of vent tiles/slates required and they should be installed in accordance with the manufacturer's instructions.

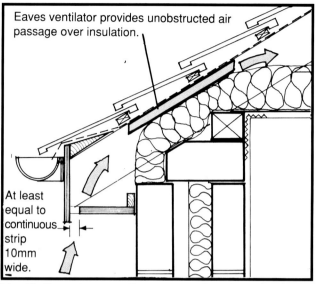

Eaves ventilator provides unobstructed air passage over insulation.

At least equal to continuous strip 10mm wide.

Ventilation at eaves the conventional method. (Provide mesh to obstruct insects etc.)

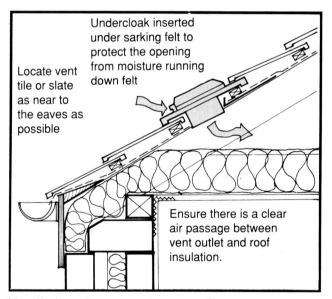

Undercloak inserted under sarking felt to protect the opening from moisture running down felt

Locate vent tile or slate as near to the eaves as possible

Ensure there is a clear air passage between vent outlet and roof insulation.

Ventilation at eaves using vent tile.

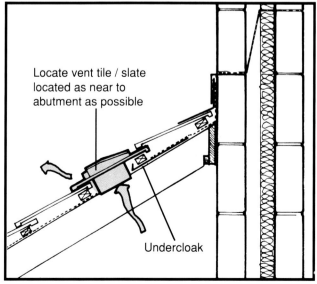

Locate vent tile / slate located as near to abutment as possible

Undercloak

Ventilation of lean to roof at abutment.

SLATE AND TILE VENTILATORS

These can be used on slated or tiled roofs where the form of construction does not allow conventional ventilation at eaves level, and in certain circumstances at mid or high level, including: parapet walls, mansard roofs, and monopitched roof construction.

Vent slate

Vent tile

How to calculate the number of tiles / slate ventilators required for a particular roof to achieve through ventilation at eaves level.

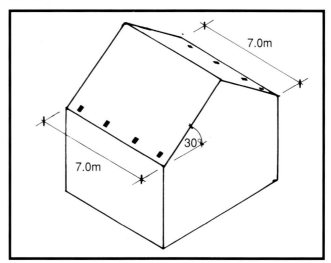

Example: The slated roof on the house has a total eaves length of 14m and a roof pitch of 30°. The Building Regulations Technical Guidance Document F requires a 10mm continuous opening over the entire eaves length (or equivalent) for a roof with a pitch of 15° or more.

For this example it is assumed the form of construction will not permit conventional ventilation at eaves, therefore proprietary slate or tile ventilators will have to be used.

Calculations

(Total eaves length) x (continuous opening size) ÷ (Capacity of vent slate/tile)=(No. of vent slates/tiles reqd.)

Note: The capacity of a vent slate / tile refers to the area it can adequately ventilate. This figure varies depending on the manufacturer. A typical figure would be 20,000mm². (Manufacturer's guidance should be sought).

$(14,000mm) \times (10mm) = (140,000mm^2)$
$140,000mm^2 \div 20,000mm^2 = 7$ no. vent. slates required min. Provide 8, i.e. 4 at each side of roof.

FLAT ROOF:
COLD DECK CONSTRUCTION

If this form of construction is used the plasterboard in the ceiling should be foil backed or a vapour barrier (e.g. 500 gauge polythene) should be tacked to the underside of the joists to control moisture build up in the insulation.

As advised on page 176 this form of roof construction should be avoided if at all possible due to the high risk of leaking, condensation and relatively low life expectancy of such construction.

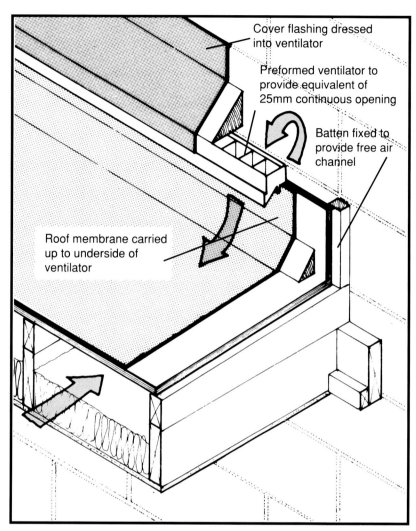

Cover flashing dressed into ventilator

Preformed ventilator to provide equivalent of 25mm continuous opening

Batten fixed to provide free air channel

Roof membrane carried up to underside of ventilator

Flat roof (cold deck construction) ventilation at abutments.

Another means of ventilating flat roofs is the use of proprietary vents of the type illustrated – manufacturer's brochures will advise on the required frequency of such vents in a roof. The type illustrated should be installed at 4.5m maximum centres.

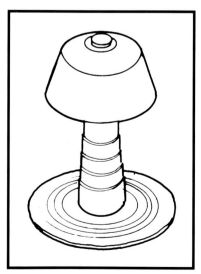

Typical "mushroom" ventilator.

DUO PITCHED ROOF WITH CENTRAL DIVISION WALL

This form of lean-to construction requires careful attention to the detail of the fixing of the head of the rafter to ensure rigidity, while maintaining the integrity of the party wall from the point of view of fire safety and sound insulation. The sketches on this page illustrate a recommended approach to such details. Supports to rafters and ceiling joists are omitted for clarity. All timber sized in accordance with SR11.

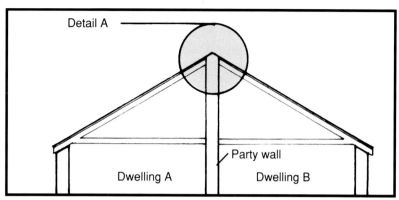

Typical "back-to-back".

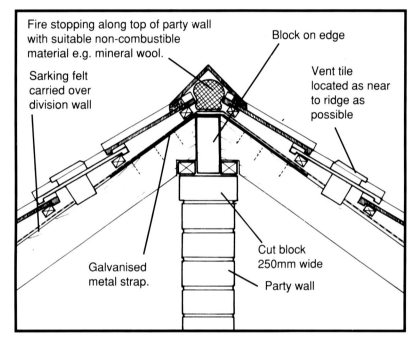

Detail A – Positive fixing of rafters in this area is vital to avoid horizontal thrust at eaves level.

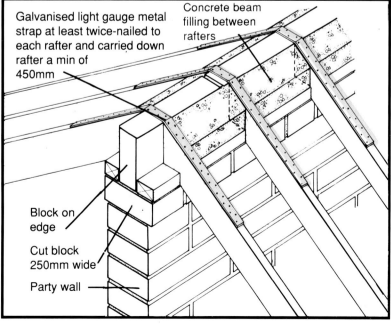

Detail A

LEAN TO ROOFS
ABUTMENT DETAIL

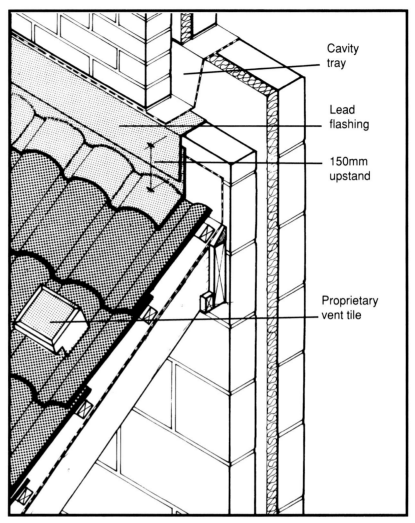

Cavity
tray

Lead
flashing

150mm
upstand

Proprietary
vent tile

Abutments of porch and lean-to roofs should be weathered with flashings built into cavity tray as shown. Ventilation to the roof should be as described previously.

PORCH REQUIREMENTS

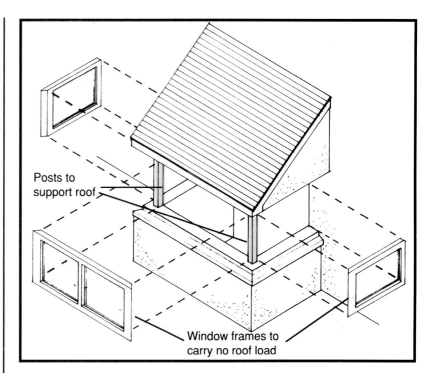

Posts to support roof

Window frames to carry no roof load

A lean to roof of the type shown here should be designed to be supported independently of the window frames.
WINDOW FRAMES MUST NOT SUPPORT ROOF LOADS.
It must be possible to remove windows entirely without affecting the roof.

Where porch or lean-to roofs are supported by posts, use a galvanised metal porch shoe to keep timber post dry.

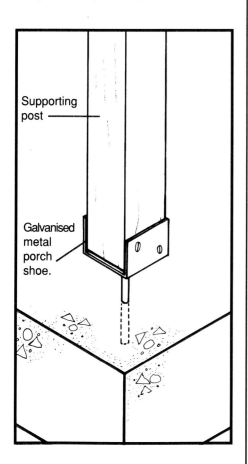

Supporting post

Galvanised metal porch shoe.

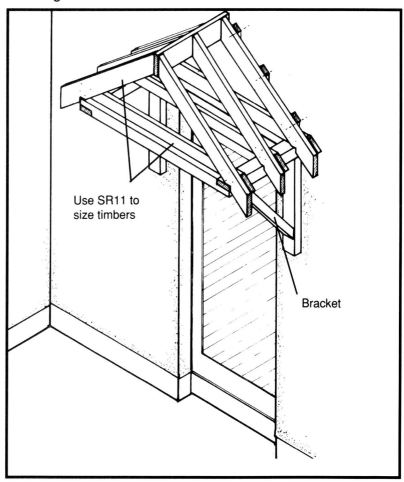

Use SR11 to size timbers

Bracket

Self supporting porch roofs should be fixed securely to walls and structurally designed.

INTRODUCTION

If at all possible, flat roof construction should be avoided due to the high risk of leaking, condensation and to the relatively low life expectancy of such construction.

Where flat roofs are used the following points should be taken into account:

◆ Concrete flat roofs without covering (to garages and outhouses for example) are prone to leaking. If a concrete flat roof is used provide a waterproof roof covering.

◆ Chipboard, including prefelted chipboard should not be used as the decking of a flat roof. Use weather and boil proof plywood (WBP) or oriented strand board (OSB)

◆ Use a good quality roof covering, make sure it has an agrément or other relevant cert. and is laid in accordance with manufacturer's instructions.

◆ Lay flat roofs so that the design fall at no point is less than 1 in 40.

◆ All water must drain away with no ponding.

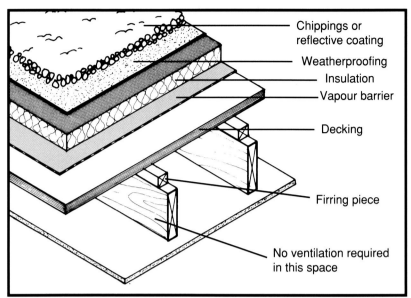

Warm deck roof: insulation above deck level, insulation must be rigid.

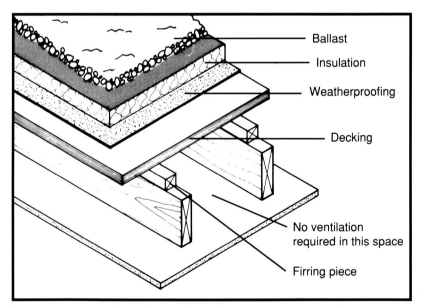

Inverted timber deck: insulation above weather proofing, insulation must be rigid and appropriate for use.

POINTS TO CONSIDER
continued.

◆ Pay particular attention at outlet points to ensure no ridges cause ponding.

◆ At balconies ensure that there is a second rainwater outlet and a 150mm upstand at any entrance doors.

◆ Provide a layer of chippings to felt roofs as a fire prevention precaution and to reduce solar heat gain. Other finishes may be treated with proprietary reflective coating.

◆ Where flat roof areas are contained by parapet and upstand walls, make provision for roof drainage overflow in the event of outlets becoming blocked.

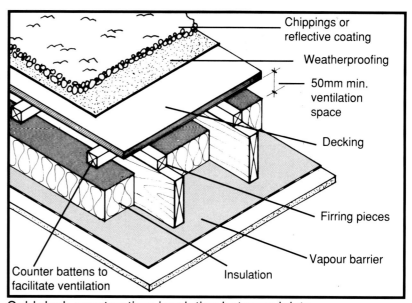

Chippings or reflective coating
Weatherproofing
50mm min. ventilation space
Decking
Firring pieces
Vapour barrier
Insulation
Counter battens to facilitate ventilation

Cold deck construction: insulation between joists.
Avoid this construction if possible.

COLD DECK CONSTRUCTION

Avoid using cold deck roof construction if at all possible. In the rare cases where it has to be used, it is necessary to ensure that the roof space is thoroughly ventilated to avoid build up of moisture and risk of timber decay.

A vapour barrier properly lapped and sealed is essential to control moisture build-up in insulation. Thorough ventilation of the roof void can be achieved by nailing counter battens on top of the firring pieces. If it is not possible to do this, an alternative means is to ventilate the roof at the abutment using preformed ventilators as illustrated on this page.

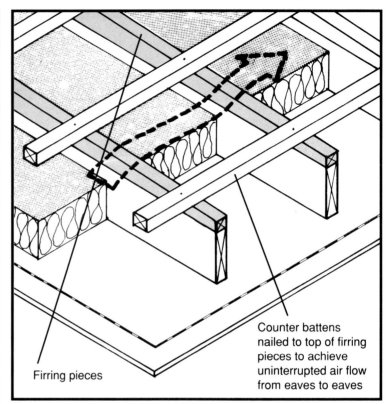

Counter battens nailed to top of firring pieces to achieve uninterrupted air flow from eaves to eaves

Firring pieces

Cold deck roof.

If this form of construction is used the plasterboard in the ceiling should be foil back and a vapour barrier (e.g. 500 gauge polythene) should be tacked to the underside of the joists, to control moisture build up in the insulation.

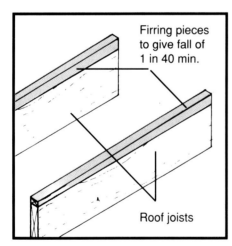

Firring pieces to give fall of 1 in 40 min.

Roof joists

Use firring pieces to achieve fall in roof (or screed to falls on concrete decks).

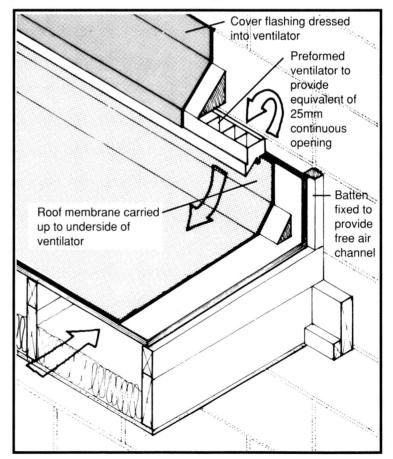

Cover flashing dressed into ventilator

Preformed ventilator to provide equivalent of 25mm continuous opening

Batten fixed to provide free air channel

Roof membrane carried up to underside of ventilator

Ventilation at abutment.

INTRODUCTION

As stated earlier on pages 114 to 117 for floor joists SR11 is an aid to the selection of appropriate sizes for structural timber. The following pages are set out, with examples of the tables from SR11 to be used in the selection of sizes for ceiling joists, rafters and purlins.

The tables in SR11 are designed to be used for sizing kiln dried timber with maximum moisture content of 22%, for normal domestic houses up to three storeys.

Use the tables in SR11 to ensure correctly sized ceiling joists, rafters and purlins.

For the design of structural timbers outside the scope of SR11 an engineer should be engaged. The engineer appointed should be qualified by examination be in private practice and possess professional indemnity insurance.

The following pages 179 to 187 explain SR11, give worked examples and reproduce the tables.

Strength Classes

SR11 divides timber in ascending order of strength into three Strength Classes (SC A, SC B and SC C) depending on the species and grade of timber.

The particular species and grades of Irish and imported timbers that fall into these Strength Classes are set out in the table below.

Soft wood Species	Strength classes		
	SC A	SC B	SC C
Irish Timber:			
Sitka Spruce	GS	SS	M75
Norway Spruce	GS	SS	M75
Douglas Fir	GS		SS
Larch		GS	SS
Imported Timber:			
Whitewood*		GS	SS
Redwood*		GS	SS
Fir-Larch**		GS	SS
Spruce-Pine-Fir**		GS	SS
Hem-fir**		GS	SS

* European ** North American

TIMBER GRADES
The grade abbreviations in the above table are as follows.

Visual grades:
General Structural: GS
Special Structural: SS

Machine grade:
M75

INTRODUCTION
continued.

It is a requirement of SR11 that all structural timber used shall be stress graded and marked accordingly. The marking system identifies the stress grade and the Strength Class of the timber and the registered number of the timber grader and the company. Both the grading and marking of timber by individual companies is subjected to the supervisory control of:

The Timber Quality Bureau of Ireland Forest Products Department, EOLAS.

Shown here are examples of the markings which occur on stress graded timber in accordance with SR11.

Span tables
The following pages contain the span tables for ceiling joists, rafters and purlins from SR11 and examples of its application. The key factor when reading the span tables is to correlate the section size and spacing of the members with the Strength Class of the timbers. It will be noted that permissible spans increase from SC A to SC B to SC C for the same section size and spacings.

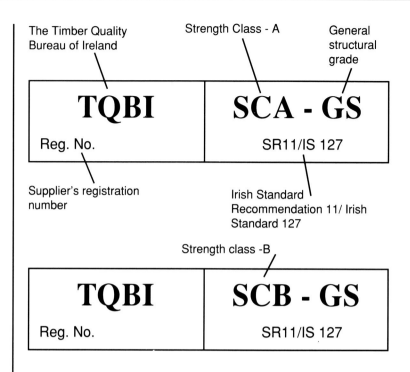

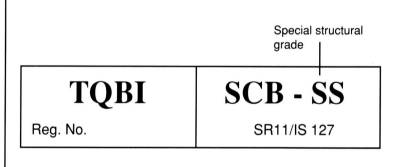

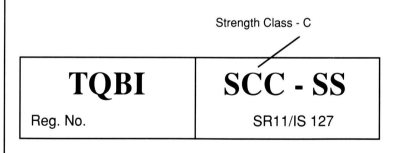

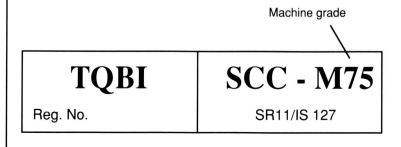

CEILING JOISTS

The permissible span according to SR11 is the clear span between supports. However as stated on page 156 traditional roof construction assumes that a binder/hanger connection supports the ceiling joists, the binder/hangers being securely fixed to the ceiling joists. Based on this assumption the span of the ceiling joist would be from the support to the binder connection as is assumed in traditional construction.

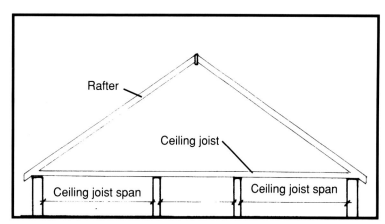

The permissible span is the clear span between supports. However traditional cut roofs construction assumes that a binder/hanger connection supports the ceiling joists, the binder/hangers being securely fixed to the ceiling joists. Based on this assumption the span of the ceiling joist would be from the support to the binder connection as is assumed in traditional construction.

Size of joist (mm)	Strength Class								
	SC A			SC B			SC C		
	Spacing of joists (mm)								
	300	400	600	300	400	600	300	400	600
	Permissible span of joists in metres								
35 x 100	1.42	1.39	1.35	1.84	1.81	1.76	1.94	1.92	1.86
35 x 115	1.82	1.78	1.71	2.24	2.21	2.14	2.37	2.33	2.26
35 x 125	2.10	2.05	1.96	2.53	2.48	2.40	2.67	2.62	2.53
35 x 150	2.88	2.79	2.64	3.26	3.19	3.07	3.45	3.37	3.24
35 x 175	3.74	3.60	3.37	4.05	3.94	3.76	4.28	4.16	3.97
35 x 200†	4.57	4.44	4.13	4.86	4.72	4.48	5.13	4.98	4.72
44 x 100	1.76	1.72	1.66	2.06	2.01	1.96	2.16	2.14	2.07
44 x 115	2.24	2.19	2.09	2.50	2.46	2.38	2.64	2.59	2.51
44 x 125	2.59	2.52	2.40	2.81	2.76	2.67	2.98	2.92	2.81
44 x 150	3.42	3.34	3.20	3.63	3.55	3.40	3.84	3.75	3.58
44 x 175	4.23	4.12	3.93	4.50	4.37	4.17	4.75	4.62	4.39
44 x 200	5.07	4.93	4.68	5.39	5.23	4.95	5.69	5.51	5.20

† This joist requires bridging at intervals of 1350mm

The above joist sizes are the minimum permissible size at 22% moisture content

Example: If a ceiling joist of strength class B is required to span 3m at spacing of 300mm. a 35x150mm, alternatively a 44x150mm joist member would be used.

ROOF RAFTERS WITH INTERMEDIATE PURLIN SUPPORT

Roof angle from 20° to 40°

The Rafter Spans shown in the tables are measured on plan and are clear spans from wall-plate to purlin or purlin to ridge. The following formula may be used to convert spans on plan to spans on slope:

Span on Slope = span on plan x K
Correspondingly:
(Span on Plan = Span on Slope ÷ K)

Roof Pitch (Degrees)	20°	25°	30°	35°	40°
K values	1.06	1.12	1.18	1.24	1.30

Intermediate roof pitch K values may be interpolated.

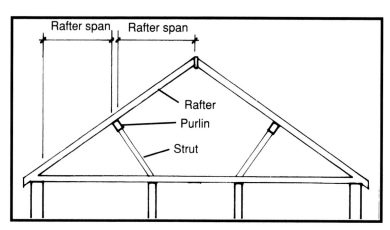

Purlin intermediate support:
The permissible span is either (1) the span between wall plate and purlin measured on plan or (2) the span between apex and purlin measured on plan.

Size of rafters (mm)	Strength Class								
	SC A			SC B			SC C		
	Spacing of rafters (mm)								
	300	400	600	300	400	600	300	400	600
	Permissible span of rafters in metres								
35 x 100	2.35	2.03	1.65	2.70	2.35	1.90	2.92	2.53	2.08
35 x 115	2.65	2.30	1.88	3.05	2.65	2.15	3.30	2.85	2.33
35 x 125	2.85	2.47	2.03	3.28	2.85	2.33	3.53	3.08	2.50
35 x 150	3.33	2.90	2.38	3.80	3.30	2.72	4.10	3.58	2.92
35 x 175	3.78	3.30	2.70	4.30	3.75	3.08	4.63	4.05	3.33
44 x 100	2.67	2.30	1.88	3.08	2.67	2.17	3.33	2.90	2.35
44 x 115	3.03	2.63	2.13	3.47	3.03	2.47	3.78	3.28	2.67
44 x 125	3.25	2.83	2.30	3.75	3.25	2.65	4.05	3.50	2.88
44 x 150	3.80	3.30	2.70	4.38	3.80	3.10	4.72	4.10	3.35
44 x 175	4.35	3.78	3.10	4.95	4.32	3.55	5.35	4.65	3.83
44 x 200	4.85	4.22	3.47	5.53	4.82	3.97	5.97	5.20	4.28

The above joist sizes are the minimum permissible sizes at 22% moisture content.

The roof rafters complying with this table are to be single length members continuous over the purlins without splices.

Example: A roof rafter of strength class B is required to span 2.5m at spacings of 400mm. A 35x115mm. member would be used.

ROOF RAFTERS WITHOUT INTERMEDIATE PURLIN SUPPORT

Roof angle from 20° to 30°.

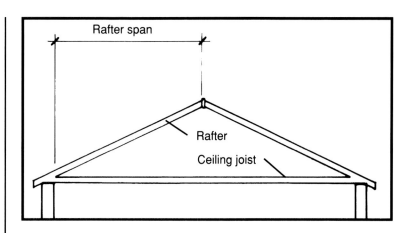

Rafter span

Rafter

Ceiling joist

No purlin support:
The permissible span is the span between apex and purlin measured on plan.

Size of rafters (mm)	Strength Class								
	SC A			SC B			SC C		
	Spacing of rafters (mm)								
	300	400	600	300	400	600	300	400	600
	Permissible span of rafters in metres								
35 x 100	1.67	1.65	1.54	1.72	1.72	1.60	1.91	1.85	1.67
35 x 115	2.02	1.92	1.82	2.16	2.07	1.95	2.21	2.15	2.06
35 x 125	2.28	2.14	2.05	2.39	2.34	2.15	2.47	2.45	2.29
35 x 150	2.88	2.72	2.43	3.04	2.85	2.69	3.21	3.04	2.79
35 x 175	3.47	3.29	2.82	3.68	3.45	3.23	3.84	3.62	3.37
44 x 100	1.84	1.76	1.67	1.97	1.89	1.79	2.11	1.97	1.89
44 x 115	2.25	2.19	2.00	2.39	2.31	2.09	2.49	2.34	2.24
44 x 125	2.46	2.34	2.29	2.66	2.52	2.33	2.80	2.69	2.51
44 x 150	3.17	3.01	2.73	3.34	3.15	2.96	3.48	3.30	3.11
44 x 175	3.77	3.68	3.16	4.01	3.86	3.51	4.25	4.00	3.65
44 x 200	4.46	4.24	3.59	4.77	4.56	4.02	5.01	4.79	4.17

The above joist sizes are the minimum permissible sizes at 22% moisture content.

Example: A roof rafter of strength class C is required to span 4.0m at spacings of 400mm. A 44x175mm member would be used.

ROOF RAFTERS WITHOUT INTERMEDIATE PURLIN SUPPORT

Roof angle from 30° to 40°.

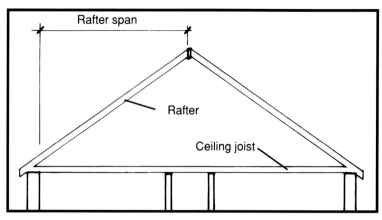

No purlin support:
The permissible span is the span between apex and wall level measured on plan.

Size of rafters (mm)	Strength Class								
	SC A			SC B			SC C		
	Spacing of rafters (mm)								
	300	400	600	300	400	600	300	400	600
	Permissible span of rafters in metres								
35 x 100	1.54	1.45	1.37	1.55	1.58	1.51	1.72	1.60	1.56
35 x 115	1.75	1.70	1.64	1.96	1.80	1.73	1.97	1.97	1.83
35 x 125	2.00	1.99	1.88	2.15	2.05	1.91	2.20	2.14	2.03
35 x 150	2.54	2.43	2.28	2.74	2.61	2.41	2.88	2.77	2.59
35 x 175	3.08	2.95	2.77	3.32	3.15	2.91	3.44	3.29	3.04
44 x 100	1.68	1.63	1.48	1.77	1.73	1.57	1.88	1.79	1.66
44 x 115	1.96	1.93	1.79	2.14	2.01	1.95	2.21	2.12	2.08
44 x 125	2.24	2.16	2.06	2.39	2.30	2.16	2.50	2.44	2.22
44 x 150	2.79	2.69	2.53	3.00	2.86	2.65	3.10	2.98	2.77
44 x 175	3.43	3.3	3.10	3.60	3.51	3.23	3.79	3.61	3:42
44 x 200	4.06	3.91	3.56	4.29	4.06	3.80	4.48	4.34	3.95

The above joist sizes are the minimum permissible sizes at 22% moisture content.

Example: A roof rafter of strength class A is required to span 3.0m at spacings of 300mm. A 35x175mm member would be used.

**PURLINS OF STRENGTH
CLASS A**

Roof angle: from 20° to 40°.

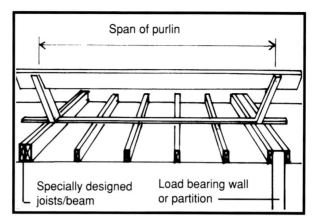

Span of purlin

Specially designed joists/beam

Load bearing wall or partition

Span of purlin: The permissible purlin span is the clear span between supports.

Purlin size	SC A								
	Span* of roof rafters in metres								
	1.25	1.50	1.75	2.00	2.25	2.50	2.75	3.00	3.25
	Permissible purlin span in metres								
75 x 225	3.16	3.12	2.82	2.63	2.56	2.40	2.35	2.21	2.17
75 x 175	2.63	2.42	2.24	2.17	2.02	1.87	1.73	1.59	1.56
75 x 150	2.26	2.17	1.90	1.84	1.59	1.55	1.42	1.39	1.36
63 x 225	3.03	2.80	2.61	2.43	2.37	2.21	2.16	2.02	1.89
63 x 175	2.43	2.23	2.05	1.89	1.84	1.60	1.56	1.53	1.40
63 x 150	2.17	1.89	1.72	1.57	1.53	1.39	1.36	1.33	1.30

The above purlin sizes are the minimum permissible sizes at 22% moisture content.

*Roof rafter span measured on plan.

Example: A purlin is required to span 2.2m and is carrying roof rafters with a 1.75m span. A 75x175mm member would be used.

PURLINS OF STRENGTH CLASS B

Roof angle: from 20° to 40°.

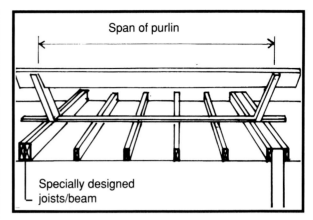

Span of purlin: The permissible purlin span is the clear span between supports.

	SC B								
	Span* of roof rafters in metres								
Purlin size	1.25	1.50	1.75	2.00	2.25	2.50	2.75	3.00	3.25
	Permissible purlin span in metres								
75 x 225	3.22	3.26	3.14	3.04	2.86	2.79	2.63	2.58	2.43
75 x 175	2.64	2.61	2.41	2.34	2.27	2.12	2.07	2.03	1.89
75 x 150	2.34	2.23	2.05	1.98	1.93	1.78	1.74	1.60	1.57
63 x 225	3.07	3.12	2.91	2.82	2.65	2.58	2.43	2.38	2.24
63 x 175	2.62	2.40	2.31	2.14	2.08	2.03	1.89	1.85	1.72
63 x 150	2.24	2.04	1.96	1.90	1.75	1.61	1.57	1.54	1.41

The above purlin sizes are the minimum permissible sizes at 22% moisture content.

*Roof rafter span measured on plan.

Example: A purlin is required to span 2.5m and is carrying roof rafters with a 3.00m span. A 75x225mm member would be used.

PURLINS OF STRENGTH CLASS C

Roof angle: from 20° to 40°.

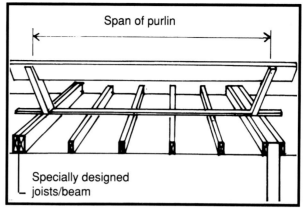

Span of purlin

Specially designed joists/beam

Span of purlin: The permissible purlin span is the clear span between supports.

Purlin size	SC C								
	Span* of roof rafters in metres								
	1.25	1.50	1.75	2.00	2.25	2.50	2.75	3.00	3.25
	Permissible purlin span in metres								
75 x 225	3.38	3.41	3.37	3.16	3.07	2.90	2.83	2.78	2.63
75 x 175	2.86	2.72	2.62	2.53	2.36	2.30	2.25	2.11	2.07
75 x 150	2.55	2.43	2.24	2.17	2.01	1.95	1.91	1.87	1.74
63 x 225	3.31	3.26	3.13	3.03	2.85	2.78	2.62	2.57	2.42
63 x 175	2.73	2.61	2.51	2.33	2.27	2.11	2.06	2.02	1.89
63 x 150	2.44	2.23	2.04	1.98	1.92	1.87	1.73	1.60	1.57

The above sizes are the minimum permissible sizes at 22% moisture content.

*Roof rafter span measured on plan.

Example: A purlin is required to span 2.5m and is carrying roof rafters with a 2.25m span.
A 63x225mm member would be used.

INTRODUCTION

A number of factors such as cost, location, aesthetics, etc., will influence whether tiles or slates are used to cover a roof. Whichever are used, the fixing recommendations of the manufacturer should be adhered to.

The details given on this and the following pages are for general guidance and normal roof construction.

Additional information on roof tiling and slating is available from:

ICP 2: 1982: Code of Practice for Slating and Tiling.

BS 5534: Part 1: 1990: Code of Practice for Slating and Tiling.

BS 5250: 1989: Control of Condensation in Buildings.

BS 8000: Part 6: 1990: Workmanship on Building Sites. Part 6 – Code of Practice for Slating and Tiling of Roofs and Claddings.

Sarking felt:
It must always be remembered that the sarking felt below tiles or slates is the second line of defence in excluding water from penetrating the roof. Care with the installation of sarking felt is therefore very important to ensure that this second line of defence is not breached.

Sarking felt should be of an appropriate type i.e. it should be "Type 1F" to I.S. 36 or BS 747.

Laps in sarking felt.
◆ Vertical laps (i.e. parallel with the rafters) should be at least 100mm and occur over a rafter so that fixings are at least 50mm from the ends of the rolls of felt being lapped.

◆ Horizontal laps (i.e. laps parallel with the battens) should be 150mm and should occur under a batten as illustrated.

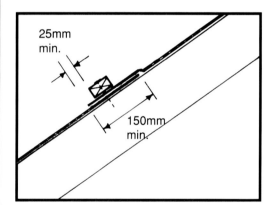

◆ At ridge level, the sarking felt should be carried over the ridge, keeping the recommended lap of 150mm.

Sarking felt at eaves.
Ensure sarking felt at eaves level is dressed 50mm min. into the gutter, and provide tilting fillet to avoid a water trap behind the fascia board. See page 196.

SARKING FELT
continued.

Sarking felt at hips.
At hips an extra layer of felt at least
600mm wide should be provided.
Lay the main felt around the hip
and the extra layer up the hip.

Sarking felt at hips.

Battens

The battens should be set out so that the
spacing between all battens is equal and does
not exceed that recommended by the tile/slate
manufacturer for the tile/slate being used and
the pitch of the roof being covered. This will
ensure that the minimum recommended headlap
is achieved for the tile/slate being used.
Minimum headlap can vary for different tile
profiles and for slates, depending on the pitch of
the roof.

Batten fixing

Battens should be at least 1.2m long (1.8m long
for trusses at 600 centres) and long enough to
be nailed at each end and at intermediate points
to at least 3 rafters. Battens should not be
cantilevered, or spliced between supports.

For duopitch roofs of not less than 17.5° with a
ridge height not exceeding 7.2m and where
basic wind speed does not exceed 48m/sec.,
battens should be nailed with 3.35mm diameter
nails, the length of the nails should be from 32 to
45mm longer than the thickness of the batten. In
more severe conditions circular ring shank nails
of the same diameter should be used. In coastal
areas, steel batten nails should be hot–dip
galvanised.

Batten ends should be cut square and butt joints
must occur over rafters. Toe nail battens to
rafters on either side of the joint. Not more than
one batten in four should be joined over any
truss or rafter.

SLATES
General information

Pitch:
Min. pitch 25° (for pitches below this the manufacturer's advice should be sought).

The steeper the pitch, the more rapid the water run–off and the better the defence against exposed weather conditions.

Batten size:
50 x 25mm for rafters up to 450mm centres
44 x 35 for rafters up to 600mm centres.

Batten fixing:
As per page 189.
Generally slates are designed to be centre nailed with 2 No. 32mm x 11 gauge copper slate nails or 32mm x 11 gauge aluminium slate nails conforming to I.S. 105 and BS 1202 and with one tail crampion or disc fitted to restrain the slate against wind uplifts.

Where exceptional local atmospheric conditions arise which could be harmful to copper, round wire silicone–bronze nails should be used. Generally, slate nails should be 20 to 25mm longer than two thicknesses of slate. Longer nails may be required at the eaves course.

Laps:
The end lap for slates will be determined by the roof pitch, the size of the slate and the angle of creep*. Manufacturer's recommendations on end lap must always be followed; where roof pitches are below 30°, extra care must be taken to ensure an adequate end lap and sub–roof to meet local conditions.

*While rain water moves down the roof to the gutter in the most direct path, there is a tendency, aided by capillary attraction, for the water to move across under the slate at the side lap. The extent of this water penetration can be described as the "angle of creep".

Hips, ridges, etc.:
A range of purpose made ridge and hip cappings and stopends is available and manufacturer's guidance on their fixing should be followed.

Flashings and soakers:
Care should be exercised when installing flashings to prevent damage to the slates. Materials for flashings and soakers should preferably be lead to BS 1178. Flashings and soakers should be code 4 and 3 lead respectively (see page 198) and gutters should be code 4 or 5 lead (see page 198).

The lead should be treated with patination oil to prevent staining of the slates.

To avoid the risk of bimetallic corrosion there should be no contact between different metals.

Bellcast eaves:
This is a common detail on a large number of houses and care should be taken in the slating of the bellcast portion to ensure adequate support for the slates. If the pitch of the bellcast is below the minimum recommended, the problem of water penetrating under the slates arises. This detail should therefore be avoided or roof pitches revised so that the bellcast portion has a pitch not less than that recommended by the slate manufacturer.

Vertical tiling:
For slates used as vertical cladding the manufacturer's recommendations should be adhered to.

SLATES
continued.

Exposure:
When designing slated roofs, account should be taken of local exposure conditions. The wind driven rain map, below, shows the range of conditions encountered in Ireland. Local conditions will influence choice of roof pitch, slate size and an acceptable end and side lap.

Generally slated roofs in the areas with an annual mean driving–rain index of greater than $7m^2/s$ will require greater laps than those required in areas of less than $7m^2/s$.

Manufacturer's guidance on recommended laps for various exposed locations should be followed.

Map of annual mean driving–rain index at a standard height of 10m above open level country.

Site laying procedure:

(1) first under–eaves course.
(2) second under–eaves course.
(3) full slate course.
(4) slate–and–a–half for use at verges, hips and valleys.

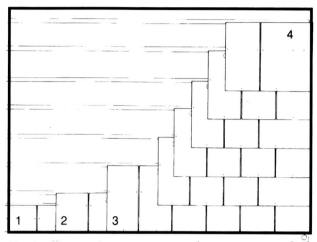

Site laying procedure

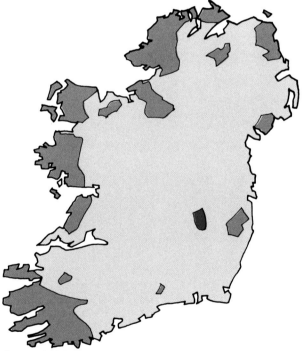

Annual driving rain index m^2/s
exposure gradings

 Over $7m^2/s$

 $3–7m^2/s$

 Under $3m^2/s$

SLATES
continued.

Reproduced opposite is an HomeBond advice note on fibre cement slates.

Fibre Cement Slates

HomeBond advise members that when using fibre cement slates problems associated with "cupping", "bowing" or "cracking" of slates may occur if slates are not fixed correctly, e.g., overtighting of crampions and/or nails, nails not driven tight enough and left proud, incorrect nails being used and incorrect spacing of battens.

HomeBond therefore advise members that if they are using fibre cement slates that they check exact fixing details with the manufacturer and that they employ specialist contractors who have the adequate skill and knowledge to ensure that slates are fitted correctly.

TILES
General information

Pitch:
Min. pitch generally 25° (however this may be reduced to 17.5° for certain types of tile if they are clipped. Manufacturer's advice should be sought).

Max. pitch generally 55° (however certain types of tiles may be fixed vertically up to 90°, manufacturer's advice should be sought).

Batten size:
35 x 35mm or 50 x 25mm
for rafters up to 450mm centres.

44 x 35mm for rafters up to 600mm centres.

Batten fixing:
As per page 189.

Tile fixing:
See also "exposure" for tile fixing specifications according to conditions.

Exposure:
The satisfactory performance of roofs depends upon the complementary function of the roof tiles and the roofing underlay.

One of the most important environmental factors which affects the satisfactory performance of roofs is wind gusting. It is during these wind gusts that pressures are set up between the roof space and the outside of the roof tiles. The result is a wind force which can cause the total or temporary removal of the roof tiles, allowing further damage by natural elements. This wind force results in a suction on both windward and leeward sides of the roof. This suction or uplifting force, particularly on a low pitched roof, is often the most severe wind load experienced by any part of a building. Under strong wind gusts, the uplift of the roof tiles may be in excess of the dead mass of the tiles, hence requiring them to be securely fixed, to prevent them from being lifted from the building.

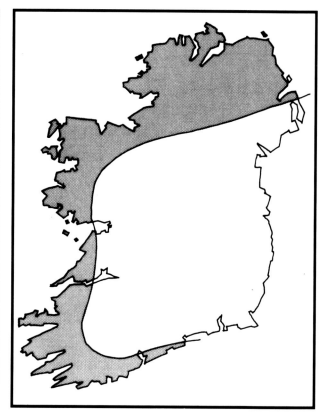

Map showing country divided into two zones on the basis of windspeeds.

Note: This map should not be interpreted literally. For locations near the boundary between the zones, fixing provisions should be based on reasoned judgement or on advice from the Meteorological Service.

TILES
continued.

Tile fixing:
Recommended Nailing/Clipping as per requirements given in Irish Code of Practice for pitched roofs ICP 2: 1982.

In the shaded area of the map on the previous page every tile should be nailed or clipped.

In the unshaded area, every alternate tile should be nailed or clipped. The nailed tiles should be staggered between rows. Every edge or perimeter tile should be nailed. In the case of valleys, the first full tile should be nailed or clipped.

On exposed sites and in built up areas subject to adverse wind effects such as funnelling, every tile should be nailed or clipped within the unshaded area.

On all roof at pitches 45° and over, each tile should be nailed or clipped. On all roofs at pitches of 55° and over, in addition to nailing each tile, the tail of the tile should be clipped.

The function of the clip is to bring much greater resistance to tile dislodgement than can be obtained by the traditional method of head–nailing at site locations and on roof pitches where the forces tending to dislodge tiles are at their most severe. Where roofs are subject to high wind lifting forces, the use of these specially designed tile clips is advisable.

The principal factors to be considered when deciding on the necessity for clipping tiles are:

◆ The exposure and location of the site.

◆ The height of the roof.

◆ The roof pitch.

◆ The higher wind loadings encountered at eaves and verges.

All these factors are discussed generally in the "exposure" section, and when any doubt persists the manufacturer's advice should be sought.

The types of nails suitable for fixing tiles are:

◆ Aluminium alloy nails to I.S. 105: part 1, suitable for normal use.

◆ Copper nails to I.S. 105: Part 1, also suitable for normal exposure.

◆ Silicon bronze nails are more durable and may be used in aggressive atmospheres.

◆ Stainless steel nails are suitable for all conditions.

◆ Galvanised iron or steel nails must not be used for tiling.

Nail sizes can be obtained from the tile manufacturer.

The clips used should be as supplied by the tile manufacturer.

TILES
Continued.

Laps:
Headlaps required for tiles depend on:

(i) the type of tile being used and (ii) the pitch of the roof. Manufacturer's recommendations should be followed.

Hips, ridges, etc.:
A range of purpose made ridge and hip tiles is available and manufacturer's guidance on their fixing should be followed.

It is recommended that a hip iron should be provided at the end of each hip rafter.

It is good practice to edge bed ridge tiles only. No ridge tile should be solid bedded except at butt joints or ridge ends. Where solid bedding is required, the mortar should be thinned out with pieces of broken tile, to reduce the mass of bedding and thus risk of cracking due to drying shrinkage. Four–point bedding (a dab of mortar at each corner of the ridge tile) is bad practice and should not be permitted.

Flashings and soakers:
Care should be exercised when installing flashings and soakers to prevent damage to the tiles.

Materials for flashings and soakers should preferably be lead to BS 1178.

Flashings and soakers should be code 4 and 3 respectively (see page 198).

Gutters should be code 4 or 5 lead
(see page 198)

Other materials complying with BS 5534: Part 1: clause 15 may also be used. To avoid the risk of bimetallic corrosion there should be no contact between different metals.

Bellcast eaves:
As stated earlier this is a common detail on a large number of homes and the manufacturer's advice should be sought in this area. If the pitch of the main roof and the bellcast roof differ by more than 5° lead soakers should be used.

Vertical tiling:
For tiles used as vertical cladding the manufacturer's recommendations should be adhered to.

EAVES AND ABUTMENT DETAILS

The eaves detail of a tiled or slated roof should be built to include the following.

◆ Fascia projection above top edge of rafter to raise up eaves course of tiles or slates to the correct pitch of roof.

◆ Tilting fillet to avoid water trap behind fascia.

◆ Sufficient overhang of sarking felt to turn down into gutter 50mm min.

◆ Overhang of at least 50mm of tiles or slates over the gutter.

Flashings

The role of flashings at junctions of roofs with other elements is vital if a sound, dry roof is to result.

The details illustrated on this and the following pages illustrate good flashing practice.

Note:

Where the roof abuts a cavity wall with a brick or fairfaced block outer leaf, a cavity tray D.P.C. should be provided to discharge over the cover flashing.

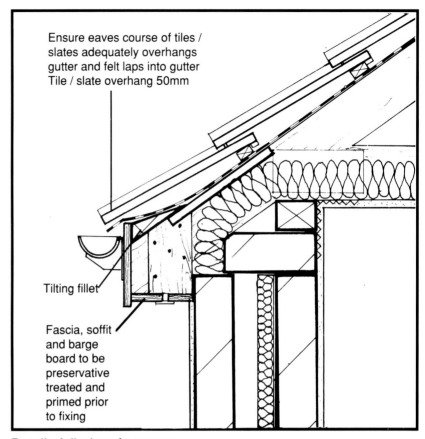

Ensure eaves course of tiles / slates adequately overhangs gutter and felt laps into gutter Tile / slate overhang 50mm

Tilting fillet

Fascia, soffit and barge board to be preservative treated and primed prior to fixing

Detail of tiled roof at eaves.

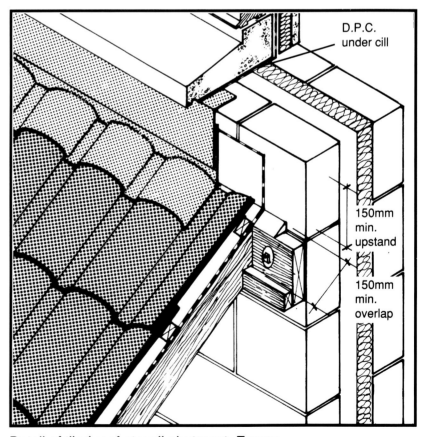

D.P.C. under cill

150mm min. upstand

150mm min. overlap

Detail of tiled roof at wall abutment. Ensure head course of tiles / slates at abutments is lapped adequately by flashing.

LEAD VALLEYS.

All leadwork on a roof should be fully supported by treated valley boards along its entire length. Lead for a valley lining should be code 4 or 5, not less than 500mm wide and in lengths of not more than 1.5m and be lapped 150mm min. at each length.

Fix treated tilting fillets on either side of the valley to provide an extra water check. Dress the lead over these fillets and welt the edges.

Fix treated tiling battens up each side of the valley. Dress the first strip of lead tightly to the boards, allowing an overhang into the gutter.

Fix the lead strip at the head with copper nails. If side fixings are used, they should only be on the upper third of the strip. Lay further strips up the valley, overlapping by at least 150mm.

Finish the roofing underlay on the valley battens. The tiling battens should be gradually swept up to the valley battens and fixed to them.

In some cases secondary valley battens, parallel to the valley battens and about 450mm away from them, may be needed to give a smooth sweep.

Lay the roof tiles into the valley and cut to rake, allowing a 100mm open channel for pitches greater than 30° and a 125mm open channel for pitches less than 30°.

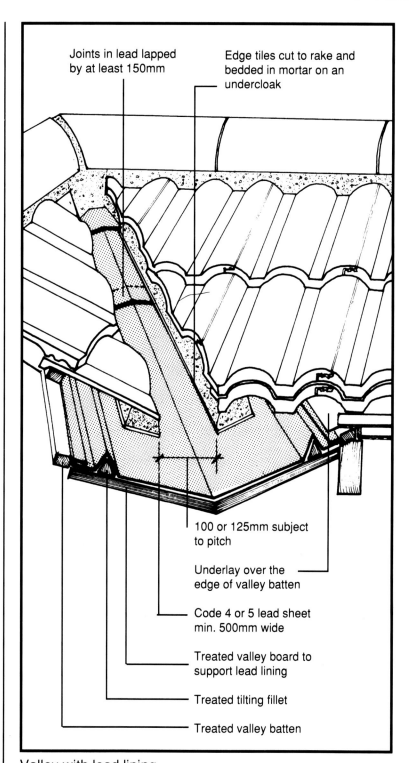

Valley with lead lining.

Joints in lead lapped by at least 150mm

Edge tiles cut to rake and bedded in mortar on an undercloak

100 or 125mm subject to pitch

Underlay over the edge of valley batten

Code 4 or 5 lead sheet min. 500mm wide

Treated valley board to support lead lining

Treated tilting fillet

Treated valley batten

LEAD VALLEYS
continued.

Lay the strip of undercloak material (natural slate or fibre cement strip is ideal) along either side of the valley to receive the mortar bedding.

Never apply mortar directly to lead as there is a high risk of movement of the lead causing the mortar to crack and break away or, worse still, splitting the lead. (Mortar mix 3:1 approx).

Apply bedding to the undercloak, making sure that a clear channel is left between the tilting fillet and the bedding. Lay the cut tiles, taking care that the sidelocks are not blocked by mortar.

Where possible, nail and/or clip all tiles on either side of the valley. Bed small tile pieces firmly in the valley mortar.

Valley checklist—
points to watch:

◆ Keep an open channel between the cut edges of the roof tiles.

1 For valley troughs below 30°, this must be 125mm min..

2 For valley troughs above 30°, it must be 100mm min..

◆ Take care that the interlocks of tiles are not blocked by mortar. This can cause damming of rainwater and leaks into the roof.

◆ Never lay roofing underlay below a lead valley. Heat from the sun may soften the bitumen and cause the lead to stick to the valley boards. There is then a high risk of the lead splitting when it expands in hot weather.

◆ Never apply mortar directly to lead. This could cause early failure of the valley.

◆ Nail and/or clip all tiles on either side of the valley. Bed small tile pieces firmly in mortar.

◆ With valley troughs, notch the fascia to ensure that the eaves tiles are not kicked up by the first valley trough.

◆ Use the proper code of lead, either code 4 or code 5. (The table below distinguishes between the codes).

Milled sheet lead to BS 1178 – 1982

Code	Colour	Thickness (mm)	Weight kg/m²
3	Green	1.32	14.97
4	Blue	1.80	20.41
5	Red	2.24	25.40

Note: Lead sheet is referred to by code, the code of the lead is its imperial designation by weight i.e. code 3 lead = 3lb/sq.ft., code 4 lead = 4lb/sq.ft. etc..

BACK GUTTER

A gutter should be formed where the bottom edge of tiling or slating meets an abutment. The gutter should be formed before tiling/slating but after felting and battening is complete.

A treated timber layboard should be provided to give support to the lead lining, the layboard should be carried up as far as the first tiling batten.

Fix a treated tilting fillet, at least 150mm wide, close to the abutment to flatten the pitch of the lead. Dress a sheet of code 5 lead into position on the timber supports. This lead should be the width of the abutment plus 450mm. The vertical upstand should be at least 100mm up the abutment. Dress the extra width of lead around the corner of the abutment after any side abutment weathering has been fitted. Dress the upper edge of the lead over the bottom tiling batten and turn it back to form a welt. Chase the abutment (25mm deep chase) insert a cover flashing of code 4 lead and dress it over the vertical upstand of the gutter.

Finish the roofing underlay with a 100mm lap onto the lead. Lay the tiles/slates in the normal way, making sure that the bottom course is not kicked up by the flat section of gutter.

In brickwork abutments adequate cavity trays must be installed to discharge over flashing.

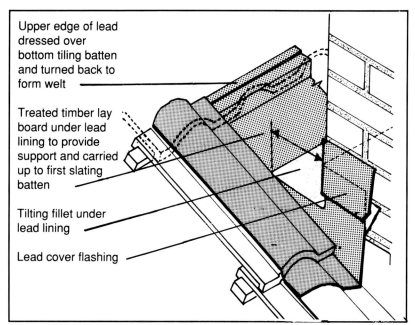

Upper edge of lead dressed over bottom tiling batten and turned back to form welt

Treated timber lay board under lead lining to provide support and carried up to first slating batten

Tilting fillet under lead lining

Lead cover flashing

Back gutter.

Abutment checklist – points to watch

◆ Lead, particularly the heavier codes, can be difficult to work. In cold weather take care not to split or puncture it whilst working.

◆ Always use proper lead working tools. Hammers are not recommended for dressing lead.

◆ Nail and/or clip all tiles next to an abutment.

◆ Fillets of mortar are not recommended at abutments because cracking of the mortar brings a high risk of failure.

SIDE ABUTMENTS

Abutment flashing with soakers.

The length of a soaker is determined by the length of the tile or slate onto which it is being fitted; and it equals the gauge of the tile or slate (i.e. the length of the tile or slate minus the lap divided by two) plus the lap (i.e. the distance one slate or tile overlaps the course next but one below it). An extra 25mm is added so that the soaker can be turned down over the top of the tile. This prevents the soaker slipping out of position. The width of a soaker should be a minimum of 175mm to allow for a 75mm upstand against the wall and 100mm under the tiles or slates. A suitable thickness of lead sheet for soakers is code 3.

Chimney flashings.

Flashings to chimney stacks which penetrate pitched roofs consist of side flashings, front apron, back gutter and metal tray D.P.C. (built into the stack in the case of brick chimneys) with integral lead flashing dressed down over the front apron.

The order of fixing to the completed stack is: first, front apron, then, side flashings and lastly, the back gutter. The metal tray D.P.C. is incorporated into the stack as it is being built.

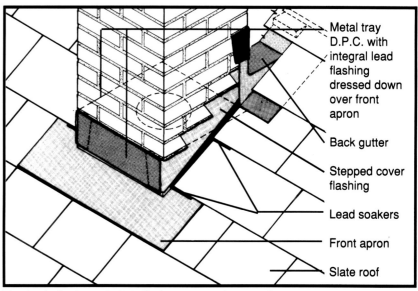

Metal tray D.P.C. with integral lead flashing dressed down over front apron

Back gutter

Stepped cover flashing

Lead soakers

Front apron

Slate roof

Typical abutment details incorporating soakers.

Front apron.

The piece of lead sheet for a front apron will be, in length, the width of the stack plus the extension beyond the stack each side for which 150 to 200mm will be needed to match the cover of the soakers or side flashing which varies according to the form of the roof covering. In width the piece of lead will need to allow 150mm for the upstand and 150 to 200mm for the cover down the roof slope; the lower the pitch, the greater the cover. Where a front apron is to be fitted to a chimney stack that has small secret gutters at the sides, the extension beyond the corners will need to be increased; the extra length required depends on the width and depth of the gutter.

SIDE ABUTMENTS
continued.

There are two common ways of weathering a side abutment with interlocking tiles:

1 Stepped cover flashings.
2 Secret gutters.

Stepped cover flashing
Turn the roofing underlay about 50mm up the abutment. Finish the tiling battens as close to the abutment as possible. Lay the tiles to butt as close as possible to the wall. Usually this can be done without cutting any tiles, by adjusting the tile shunt across the roof. Half tiles may be useful. Cut a piece of code 4 lead as shown to form a combined step and cover flashing. The flashing should not be more than 1.5m long and should be wide enough to cover the abutting tile by 150mm, or to cover the first roll, whichever gives the greatest cover. Chase out the brickwork mortar joints (25mm deep chase). Push the folds of the flashing into the chase and wedge in with small pieces of lead. Dress the cover flashing as tightly as possible to the tile profile. Then repoint the brickwork.

Note:
If the abutment is rendered the render should not be applied directly to the flashings, as this restricts movement and may cause splitting of the flashing or detachment of the rendering. Use proprietary expanded metal stop bead 75mm above finished roof line.

Note:
The lead flashing should be chased into the joints in the brickwork – it should not be nailed directly to the brickwork.

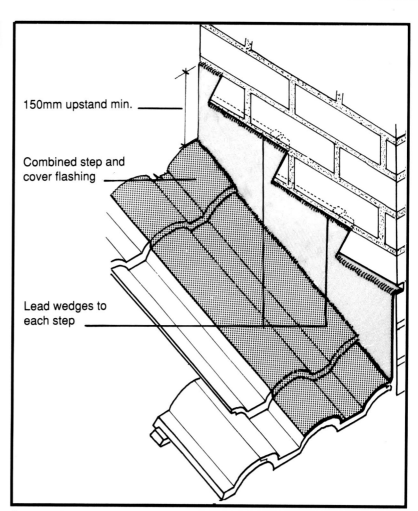

150mm upstand min.

Combined step and cover flashing

Lead wedges to each step

Stepped cover flashing on contoured tiles.

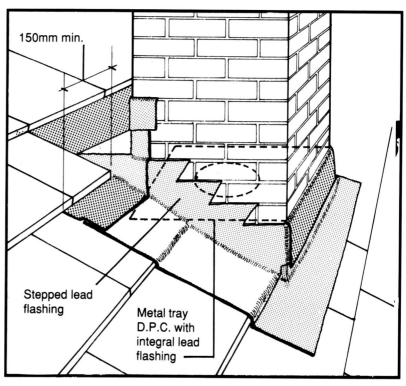

150mm min.

Stepped lead flashing

Metal tray D.P.C. with integral lead flashing

Stepped cover flashing on flat interlocking tiles.

SECRET GUTTER

The secret gutter should be formed before tiling commences.

A treated timber support should be fixed between the last rafter and the abutment. This should be at least 75mm wide and run the full length of the abutment.

Fix a splayed timber fillet at the discharge point to raise the lead lining to the right height. Take care not to create a backward fall with this fillet.

Fix a treated counter batten along the outer edge of the rafter. This should be thinner than the main tiling battens by at least two thicknesses of lead.

Line the gutter with code 4 or 5 lead, in lengths of not more than 1.5m. Lap each strip over the lower one by 150mm and fix with copper nails at the head.

The lead welts should be turned up to exclude birds and vermin from entering the tile batten space.

The gutter should be about 50mm deep and have a vertical upstand of at least 100mm against the abutment.

At the discharge point, the lead lining should be splayed out across the timber fillet to avoid the risk of an overflow at the side.

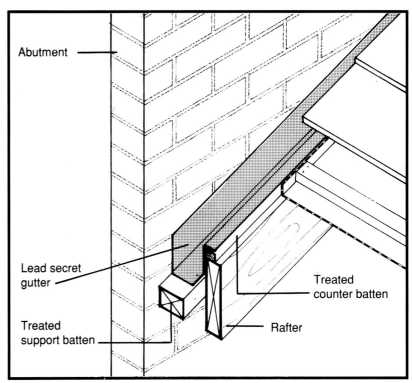

Typical section through lead secret gutter.

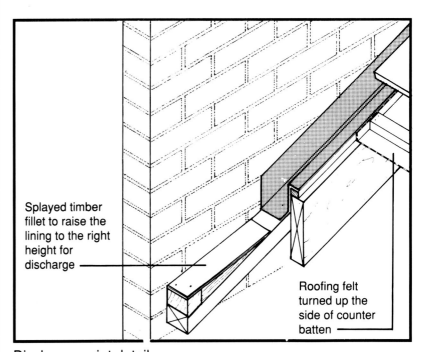

Discharge point detail.

SECRET GUTTER
continued.

Turn the roofing underlay up the side of the counter batten. The tiling battens should butt up to the counter batten.

Lay tiles to leave a gap of 25 – 38mm by the abutment. This allows for future cleaning out of the gutter to avoid blockages.

A cover flashing above the secret gutter is advisable, particularly in areas of high exposure or on roofs under trees where the risk of blockages is high.

Fit a stepped flashing chased into the brickwork (25mm deep chase) and dressed over the vertical upstand.

Note:
If abutment is rendered the render should not be applied directly to the flashings as this restricts movement and may cause splitting of the flashing or detachment of the rendering. Use proprietary expanded metal stop bead 75mm above finished roof line, see detail on page 69.

Where the roof abuts a cavity wall with a brick or fairfaced block outer leaf, a cavity tray D.P.C. should be provided to discharge over the cover flashing.

In brick and fairfaced block chimneys, a metal tray D.P.C. should be similarly incorporated. See pages 67 – 70 and 110.

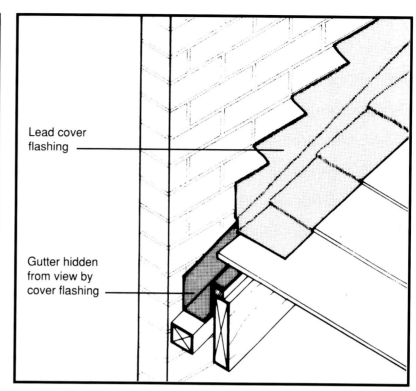

Complete secret gutter with lead cover flashing

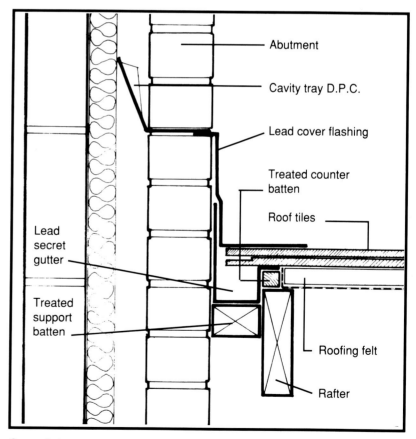

Complete secret gutter

FLAT-ROOFED DORMER WINDOW

Where pitched roofs incorporate dormer windows which have flat roofs the following recommendations apply:

◆   Where rainwater discharges from the flat roof of a dormer window onto a pitched roof a lead apron should be provided at the point of discharge as illustrated.

◆   Where only narrow widths of pitched roof occur adjoining flat–roofed dormer windows, the rainwater runoff from the roof should be arranged so as not to discharge onto such areas.

◆   The drainage fall on the flat roof should be towards the vertical face of dormer and not inwards towards the junction of flat roof and pitched roof.

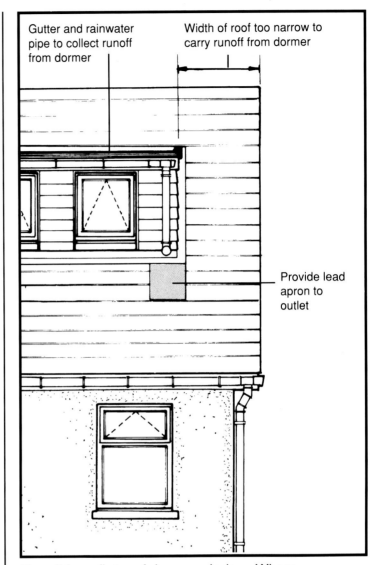

Gutter and rainwater pipe to collect runoff from dormer

Width of roof too narrow to carry runoff from dormer

Provide lead apron to outlet

Runoff from flat roof dormer window. Where possible rainwater should not discharge over roof.

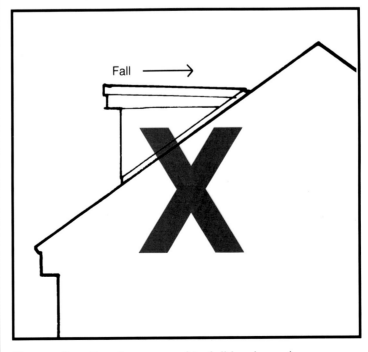

Fall →

Never allow the dormer roof to fall backwards.

SIZING GUTTERS AND DOWNPIPES

To efficiently size gutters and downpipes the effective area of the roof being drained must first be established. This depends on the pitch of the roof, see table 1 opposite. When the effective area of the roof has been calculated and the location of the downpipes determined, the size of the gutter and downpipe can be obtained from the graph below, (see examples on following page).

The guidance on these pages has been derived from Building Regulations Technical Guidance Document H and BS 6367: 1983: Drainage of Roofs and Paved Areas and applies only to straight lengths of gutters. Where a gutter includes an angle reference should be made to BS 6367 for guidance.

Table 1	
Roof type	Effective area
Flat 30° Pitch 45° Pitch 60° Pitch 70° Pitch or greater	Plan area Plan area x 1.15 Plan area x 1.40 Plan area x 2.0 Elevational area x 0.5

Gutter and outlet size graph. (half-round gutters).

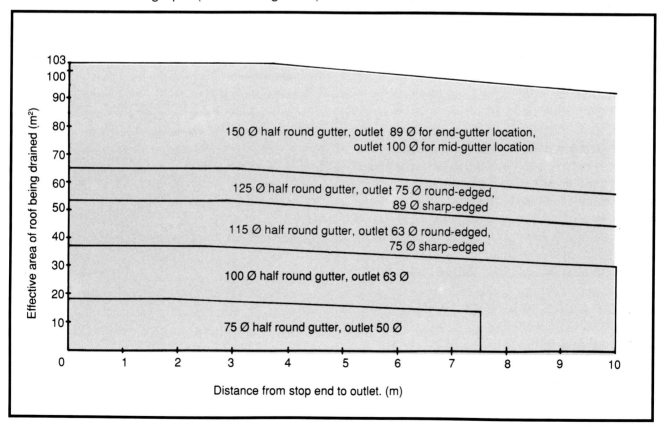

Notes 1. Gutters and downpipes may be omitted from a roof at any height provided that it has an area of 6m² or less and provided no other roof drains onto it.
2. Rainwater pipes for standard eaves gutters should have the same nominal bore as the gutter outlets to which they are connected.

SIZING GUTTERS AND DOWNPIPES

Example 1

Detached house and garage

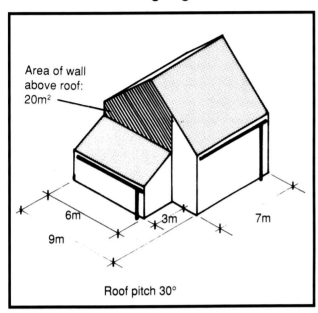

Area of wall above roof: 20m²

6m 3m 7m
9m

Roof pitch 30°

Example 2

Semi–detached houses

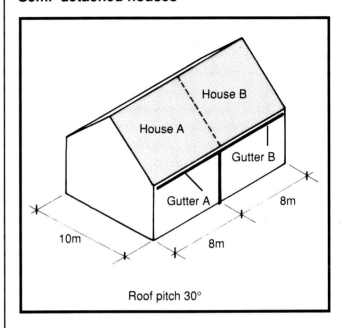

House B
House A
Gutter B
Gutter A
10m 8m 8m

Roof pitch 30°

1 **Garage:** 3m x 6m (plan area of garage roof) x 1.15 (multiplying factor for 30° roof pitch from table 1) = 20.7m²

20m² (area of wall above roof) x 0.5 (multiplying factor for surfaces of 70° pitch or greater from Table 1) = 10m²
Effective roof area:
20.7m² + 10m² = 30.7m²

From the gutter sizing graph on the previous page it can be seen that a 100mm dia. half round gutter with 63mm dia. outlet is sufficient for this area of roof.

2 **House:** 7m x 4.5m (plan area of house roof) x 1.15 (multiplying factor for 30° roof pitch from table 1) = 36.2m² (effective roof area).

From the gutter sizing graph on the previous page it can be seen that a 115mm dia. half round gutter with 63mm dia. outlet is sufficient for this area of roof.

Semi–detached houses:
Note: Roof area of house A discharges to gutter A and roof area house B discharges to gutter B, both gutters discharge to a single downpipe, located centrally.

Calculation:
8m x 5m (plan area roof of one house x 1.15 (multiplying factor for 30° roof pitch from table 1) = 46m² (effective roof area of 1 house)

From the gutter sizing graph on the previous page it can be seen that a 115mm diameter half round gutter with a 63mm. diameter round-edged outlet or 75mm diameter sharp-edged outlet is adequate for the area of roof in question.

VERGE DETAILS

The construction of roof verges should follow the guidance indicated on this page. Secure nailing of the gable ladder to the last rafter inside the gable wall is needed for the stability of this section of a roof.

Note: The cut slate is required to hold the verge pointing. **It is not acceptable to nail a batten to the barge and remove it after pointing has set.**

Gable ladder securely nailed to last truss at 400mm centres.

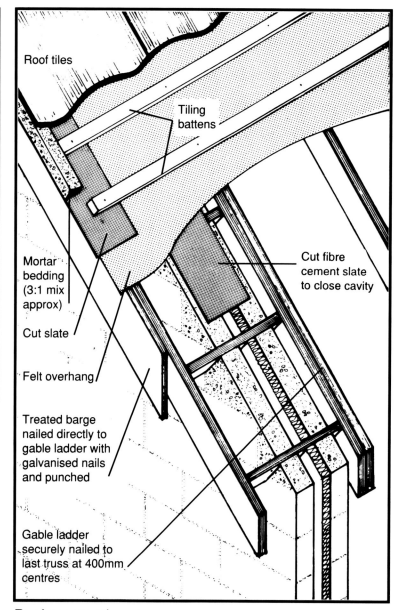

Roof tiles

Tiling battens

Cut fibre cement slate to close cavity

Mortar bedding (3:1 mix approx)

Cut slate

Felt overhang

Treated barge nailed directly to gable ladder with galvanised nails and punched

Gable ladder securely nailed to last truss at 400mm centres

Roof construction at verge.

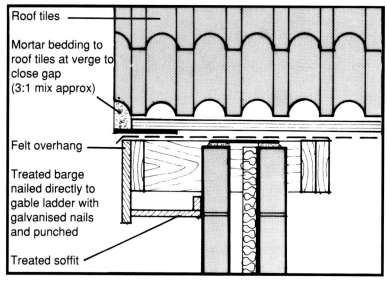

Roof tiles

Mortar bedding to roof tiles at verge to close gap (3:1 mix approx)

Felt overhang

Treated barge nailed directly to gable ladder with galvanised nails and punched

Treated soffit

Roof construction at verge (Section).

SECOND FIX, PLASTERWORK AND SCREEDS

FIXING INSULATION

For guidance on insulation thickness, see appendix K.

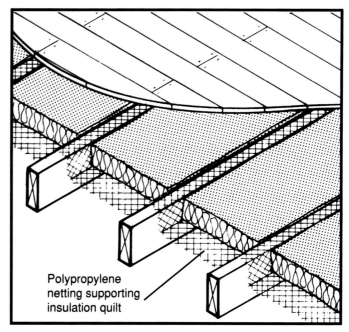

Polypropylene netting supporting insulation quilt

Suspended timber floors.
Insulation quilts should be supported by draping polypropylene netting across the joists and stapling it to joist sides so that the quilt can be laid to the full thickness.
Insulation draped over joists is not acceptable. Rigid insulation boards should be supported on battens or fillets nailed to the sides of joists.

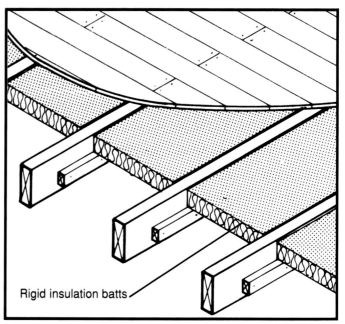

Rigid insulation batts

Rigid insulation supported by battens fixed to joists.

INTERNAL FINISHES TO BLOCKWORK

There are two types of wall finishes used. One is the use of wet plastering techniques and the other is drylining with plasterboard. The wet plastering method will require a certain amount of drying out time before decorating, whereas the drylining system may not. The type of insulation used will also be a determining factor on which system is used. With hollow block construction, the insulation will have to be placed on the inside face of the wall and dry lining will be required.

Reference should be made to the sound requirements for party walls in semi-detached or terraced houses, as Technical Guidance Document E of the Building Regulations requires that a concrete block party wall should be at least 415kg/m^2 and be plastered on each face with plaster of at least 12.5mm thickness. As defined in the Technical Guidance Document concrete blocks of 1860kg/m^3 with 12.5mm lightweight plaster each side will meet this requirement.

Wet plastering to blockwork, alternative procedures:

1   Scud blockwork to a depth of 3–5mm using a thick slurry of 1 part sand to 2 parts cement.

Apply a scratch coat to a depth of 10–16mm using a 1:1:6 mix (cement: lime: sand) or 1: 6 mix (cement: sand) with plasticiser added in accordance with the manufacturer's instructions. Scratch surface thoroughly to form a key. Finish with a 2mm coat of gypsum plaster when scratch coat has dried.

Note: If scratch coat is not allowed to dry fully, subsequent failure of gypsum plaster finish can occur as the photo on page 212 shows, or,

2   Apply a coat of at least 9mm of proprietary gypsum base coat plaster, well scratched to form a key for finishing coat. Finish with a 2mm finish coat of gypsum final coat plaster, when basecoat has set.

INTERNAL FINISHES TO BLOCKWORK
continued.

Allow basecoat to dry thoroughly, otherwise plaster finish will fail.

Getting the basecoat right is vital to prevent failure of the finishing coat.

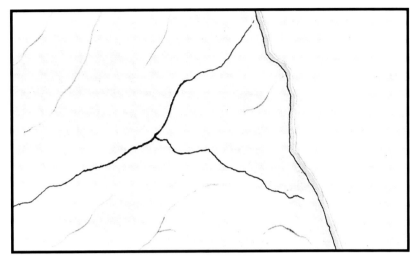

Shrinkage cracking in sand/cement basecoat causing delamination (boasting) of finished coat.

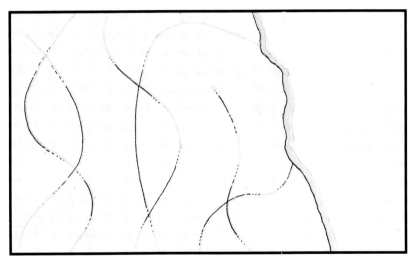

Poor and inadequate scratching – little if any key for finishing coat.

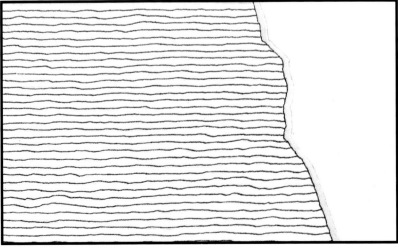

Good scratching of basecoat forming key for finishing coat.

PLASTERBOARD BASED FINISHES TO WALLS AND CEILINGS

As an alternative to plaster finish being applied to blockwork as described on the preceding pages, internal wall finishes are commonly provided by plastering methods incorporating the use of plasterboard. Plasterboard systems also form the basis of the vast majority of ceiling finishes in domestic construction.

Plasterboard based methods for wall and ceiling finishes fall into two types – "dry" systems and "wet" systems.

"DRY" PLASTERING SYSTEM

These systems incorporate the use of large plasterboard sheets, 12.5mm thick, on walls and ceilings, using joint filling methods which leave the board ready for decoration.

Walls – fixing of plasterboard

Fix boards to vertical treated battens with bound edges of boards vertical. Battens to be at 600mm maximum centres. Provide a continuous horizontal treated nogging at the top and bottom of each sheet. This is required for proper fixing of the sheet and to fire stop the junction of ceiling and wall. Such noggings should be securely fixed to the wall at 450mm centres max.. External corners such as door and window reveals should be reinforced with flex corner tape or angle bead.

Alternatively, fix plasterboard sheets to walls using dabs of drywall adhesive. Fire stopping at floor and ceiling level is also required and this can be done by using strips of drywall adhesive or timber noggings as described in the preceding paragraph.

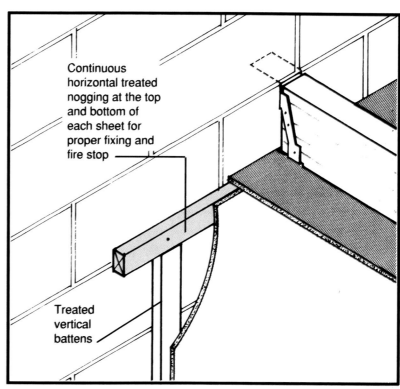

Continuous horizontal treated nogging at the top and bottom of each sheet for proper fixing and fire stop

Treated vertical battens

Wall lining fixed to treated battens.

"DRY" PLASTERING SYSTEM
continued.

In cases where insulation material is laminated to the back of plasterboard slabs and these are fixed by dabs of drywall adhesive, the floor and ceiling level should be fire stopped by the use, for example, of treated timber noggings top and bottom and the plasterboard nailed to these noggings and also mechanically fixed at each edge of the board at mid-height, through the dabs.

The recommendations of the manufacturers of plasterboard incorporating insulant backing should be followed, together with the requirements of any Agrément Certificates.

Ceilings – fixing of plasterboard

Plasterboard should be fixed lightly butted, with bound edges running at right angles to the joists and should be staggered so as to avoid a continuous end joint. Recommended spacing of joists is 450mm, which may be increased to a maximum of 600mm if noggings are fixed between joists to provide for fixing of the bound edge of the sheets. It is recommended that all edges be supported at the perimeter. Nail the boards to every support at approx. 150mm centres (8 nails per linear metre), working from the centre of the board outwards. Nails should be driven home firmly without the head fracturing the paper surface but leaving a shallow depression to facilitate spotting. For 12.5mm board nails to be 40mm long galvanised, with 7mm diameter head and 2.5mm diameter shank. Ensure that all slabs are fixed level and flat and that opes for services are formed neatly.

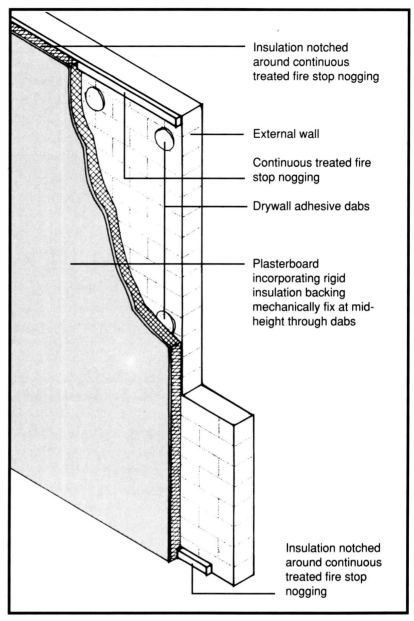

Insulation notched around continuous treated fire stop nogging

External wall

Continuous treated fire stop nogging

Drywall adhesive dabs

Plasterboard incorporating rigid insulation backing mechanically fix at mid-height through dabs

Insulation notched around continuous treated fire stop nogging

Wall lining incorporating insulation – typical fixing details.

Wall and ceiling finishes alternatives:

1   Tape and fill all joints and internal angles and leave ready for direct decoration by applying a slurry of joint finish or a proprietary sealing agent.

2   Tape and fill all joints and internal angles and finish with 2mm of skimcoat plaster finish suitable for use on drylining boards.

Note: In all cases described above the tapered edge of the plasterboard should be on the face – "IVORY FACE VISIBLE"

"WET" PLASTERING SYSTEM

This system involves the use of round edge plaster lath (standard sizes: 1200x400 and 1219x400). Standard board thicknesses are 9.5mm and 12.5mm. The table below sets out the recommended and maximum spacings of joists for each thickness of board. See also page 123 for guidance on the fixing of plasterboard in the construction of timber stud partitions.

Board thickness (mm)	Recommended centres (mm)	Maximum centres (noggings required) (mm)
9.5	400	450
12.5	450	600

Adverse site conditions (e.g. high humidity, slow drying) can cause sagging of plasterboard. The recommended centres take account of such conditions as far as possible.
The recommended centres should not be exceeded without the use of noggings.

Round edge plaster lath – joist spacing.

Ceilings – fixing of plaster lath
Plaster lath should be fixed with bound edge running at right angles to the joists and should be staggered so as to avoid a continuous end joint. It is recommended that all edges be supported at the perimeter. Nail the boards to every joist at approximately 150mm centres (8 nails per linear metre) working from the centre of the boards outwards. Nails should be driven home firmly without the head fracturing the paper surface but leaving a shallow depression to facilitate spotting.

Reinforce all internal corners, such as wall-to-ceiling junction, with 90mm jute scrim. Caulk all joints with board finish plaster. Reinforce external angles such as the corners of downstand beams with galvanised metal angle bead.

Ensure 3mm joints between bound edges of boards. Lightly butt cut edges and butt ends of boards. Nails to be 30mm long for 9.5mm board and 40mm long for 12.5mm boards, galvanised, with 7mm diameter heads and 2.5mm shanks.

Ceiling finish:
Apply a coat of bonding plaster to a nominal thickness of 8mm well scratched to form a key for finish. Finish with a 2mm nominal thickness of finish coat steel trowelled to a smooth finish.

How to avoid plaster cracks in ceilings.
1   Use properly dried timber – moisture content should not exceed 22%.

2   Ensure that all joints are properly caulked, scrimmed or taped as appropriate to the board being used.

3   Provide sufficient support, especially noggings.

4   Ensure that struts which prop purlins in cut roof construction are correctly supported.

5   Ensure that joists are correctly sized for their span – use SR11.

Wall finish:
Apply a 3mm coat of gypsum board finish plaster, steel trowelled to avoid loose patches and to form a smooth finish, ensuring all reveals are accurately formed.

To avoid damage due to dampness and/or humidity no slabbing should be done until the house is closed in, i.e., fully roofed, doors and windows fitted and sealed.

INTRODUCTION

Sand and cement screed is a common form of finishing layer on precast and in–situ concrete floor slabs. Screeds can be used in three principal ways, as described opposite: bonded, unbonded and floating screed.

For the tamped finish usual on domestic concrete floor slabs, a screed at least 50mm thick is necessary in the case of unbonded screeds to minimise the risk of curling. Where the screed is laid directly on insulation (floating screed) a minimum thickness of 65mm (it is recommended light mesh reinforcement be incorporated) is needed to avoid curling.

Screed mixes
Generally, the mix for a screed should be 1:3 cement: dry sand, wetted sufficiently to allow adequate compaction. Too dry a screed mix will make such compaction difficult.

In addition to ensuring that screeds have the minimum thickness appropriate to their type (i.e. bonded, unbonded or floating) the risk of curling of screeds can be further minimised by using proprietary bonding agents such as PVA (polyvinyl acetate) or SBR (styrene–butadene rubber) in the screed mix to replace water. Such bonding agents should be used in strict compliance with the manufacturer's recommendations.

Screed types

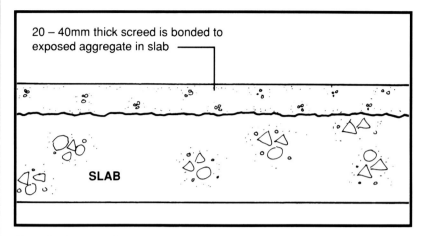

20 – 40mm thick screed is bonded to exposed aggregate in slab

SLAB

Bonded screed:
To ensure a properly bonded screed the base slab surface must be roughened (e.g. by scabbling), have all dust removed, be soaked by water and have a bonding agent or cement grout brushed well in. The screed is laid immediately after wetting and grouting of the slab and is well compacted. Bonded screeds are laid to a thickness of 20 – 40mm. If laid to a thickness of greater than 40mm, a bonded screed is more likely to curl or become hollow.

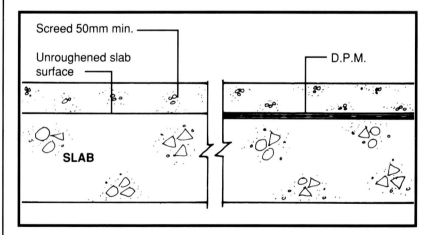

Screed 50mm min.

Unroughened slab surface

D.P.M.

SLAB

Unbonded screed:
Unbonded screed laying will allow some risk of curling and hollowness, as the unroughened base or damp proof membrane prevents bonding. However the risk of curling and hollowness can be reduced by increasing screed thickness, and therefore its weight – thus for unbonded screeds a minimum thickness of 50mm is considered necessary.

SCREEDS
continued.

A further step in reducing the risk of curling of screeds is the substitution of small–sized aggregate (i.e. 10mm) to replace a proportion of the sand in the mix. Where this is done a higher workability (around 50mm slump) will be required rather than the stiff consistency characteristic of sand/ cement screed mixes.

Curing
Cure screeds by covering with plastic sheeting immediately after laying. Lap sheets well and keep in place for at least 7 days. In addition to achieving a well cured screed, the sheeting helps protect the screed from foot traffic. Rapid curing of screeds will increase the risk of cracking and curling.

Service pipes etc. in floors

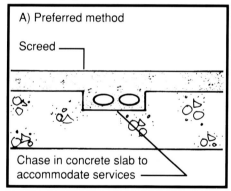

A) Preferred method

Screed

Chase in concrete slab to accommodate services

It is preferable to accommodate services pipes etc. in the floor slab. Where this is not possible and service pipes are accommodated in the floor screed the guidance below should be adhered to.

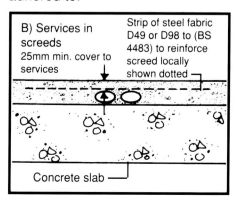

B) Services in screeds
25mm min. cover to services

Strip of steel fabric D49 or D98 to (BS 4483) to reinforce screed locally shown dotted

Concrete slab

Screed types
continued.

Screed 65mm min.

Insulation

SLAB

Floating screed:
Where screed is laid directly on a layer of insulation a minimum thickness of 65mm is considered necessary.

Lay screeds to a proper thickness.
Thin screeds are prone to lifting and cracking

Cracking of screed due to inadequate cover on pipes.

INTRODUCTION

See the detailed guidance on pages 205 and 206 in relation to the procedures for the correct sizing of gutters and downpipes to ensure adequate capacity for the area of roof being drained.

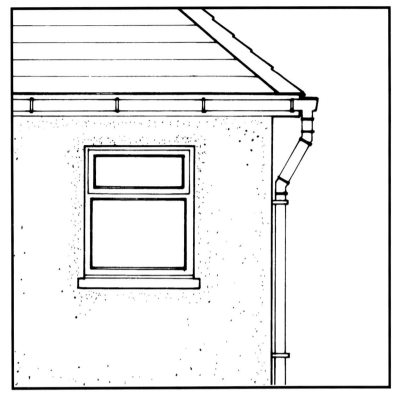

Ensure that gutters are of adequate size and laid to proper fall. Ensure adequate number of downpipes
Provide sufficient clips to gutters and drainpipes.

SEQUENCE OF OPERATION

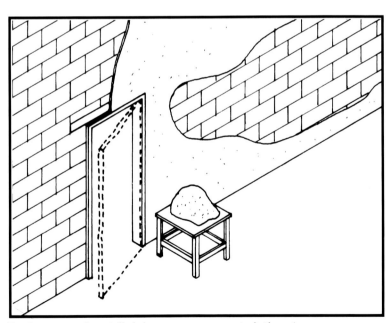

Drying out of wet finishes can cause twisting / warping of doors if hung too early.

DRYING OUT

Ensure that building is closed to the weather before commencing finishing trades, thus avoiding damage to finishes by exposure to the effects of the weather.

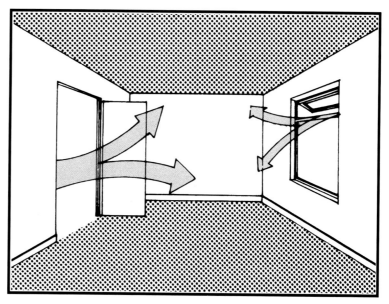

Allow adequate ventilation to speed up drying of wet finishes. It is also important to allow adequate time for drying out.

ALL STAGES

PROTECTION GUIDELINES

To avoid costly repair work, every effort should be made to protect completed work from damage and carelessness.

The following guidelines should be applied.

◆ Avoid heavy construction traffic over external door thresholds as waterbars get damaged thus decreasing watertightness in the building.

◆ Protect doors with polythene or original wrapping.

◆ Protect door frames and linings with timber strips or plywood to at least 1m above floor level.

◆ Flooring should be protected with suitable temporary covering (e.g. building paper) to withstand damp caused by plaster droppings and the like.

◆ Protect stair treads and handrails with timber strips, plywood or building paper. Avoid the use of polythene.

◆ Protect installed kitchen worktops and unit faces. Do not use worktops as work benches or cupboards for temporary storage of tools and site materials.

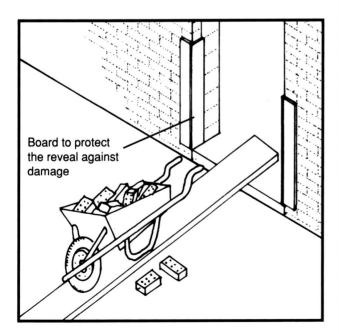

Board to protect the reveal against damage

◆ Protect reveals

TIMBER PRESERVATION

All external softwood timber components such as windows, external doors and frames, fascia boards, barge boards, etc., should receive a double vacuum preservative treatment prior to delivery to the site. A certificate of treatment should be obtained from the timber supplier. The cut ends of pretreated timber should receive a liberal brush treatment with preservative on site.

Surface coating of external joinery timbers

Paint finish:
Where joinery timbers are intended to receive a paint finish they should first be primed on all surfaces including concealed areas. This reduces moisture uptake, expansion and shrinkage in service. Ideally joinery should be shop primed. Where this is not possible ensure all joinery components are primed promptly upon delivery to the site.

Clear finish:
Where joinery timbers are intended to receive a clear surface finish they should be sealed as early as possible on all exposed and concealed surfaces with a sealing coat compatible with the intended finish.

SITE STORAGE

In the interest of cost saving and site safety appropriate measures should be taken for the protection and storage of materials on site.

The following guidelines should be applied.

Joinery:

Ensure that joinery on site is kept dry at all times. If stored in the open, cover top and sides allowing ventilation to the stack, and also keep the stack well clear of the ground.

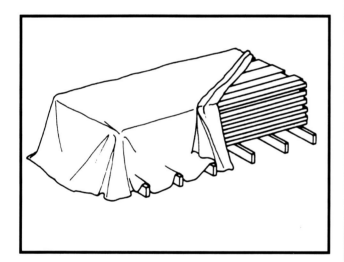

Bagged materials:

Hydrated lime, cement and premixed dry bagged materials should be stored off the ground, under cover and away from damp surfaces, in such a manner as to allow the bags to be used in rotation in order of delivery.

Ready-mixed lime : sand for mortar should be tipped onto a cleaned sealed banker board with a sealed base and should be covered when not in use.

Covering over is particularly important when coloured mortars are being used, because rain and weathering may wash out some of the pigment or other fine materials and lead to a variation in colour.

Plasterboard:

Plasterboard should be stored off the ground and horizontally on a level base consisting of a timber platform or bearers at least 100mm wide laid across the width of the boards at centres not exceeding 400mm.

If plasterboard is not stored in a weatherproof building, completely cover the stack with a waterproof sheet secured all round. Protect from damp rising from below the stack. Unless special provisions are made, do not stack boards to a height of more than 1m.

Insulation:

Insulation products should be delivered to site suitably wrapped in polythene or similar material. They should be stored under cover and protected from direct sunlight. Contact with solvents and organic based materials such as coal tar, pitch or creosote should be avoided.

Additional advice on handling and storage along with health and safety requirements is available from manufacturers.

SITE STORAGE
continued.

Trussed rafters

Trussed rafters should be:

◆ Stored and handled to prevent distortion and sagging or bending.

◆ Stored on level bearers at wall plate position, vertically for long term storage (horizontal stacking is acceptable for short term storage).

◆ Cover to protect from rain and sun, whilst ensuring good ventilation.

◆ Stored in a secure manner.

Prolonged dampness may cause rot in timber and corrosion of the connector plates.

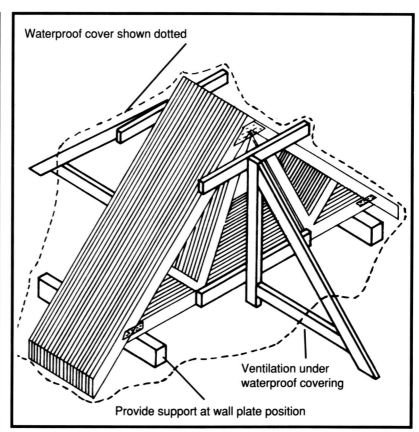

Long term storage.

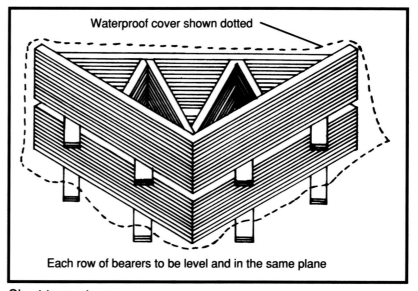

Short term storage.

ASSESS WEATHER CONDITIONS AND PLAN ACCORDINGLY

Monitor local temperatures and provide a minimum/maximum thermometer on site to determine the shade temperature. If properly used, it will show at a glance if the temperature is rising or falling.

Don't wait until the temperature is at freezing before you think about cold weather precautions. Plan the work as far as possible to avoid freezing conditions.

Temperatures in direct winter sunshine are not a reliable guide – do not refer to them, work to shade temperatures which may be much lower.

Consider the possible rapid drop in temperature after sunset.

PROTECTION OF STORED MATERIALS

Overnight protection

Where it is intended to carry out work during cold weather provide covers to all materials which are susceptible to frost.

These precautions are additional to those normally used to protect materials. The use of covers will protect materials from overnight snow, ice and frost. They will also reduce the effects of longer term frosts and permit an earlier resumption of work.

In cold weather always use waterproof coverings over bricks, blocks, sand and ballast, to prevent them from becoming saturated and affected by overnight frost.

Longer cold periods

To carry out work during prolonged cold weather provide additional protection to susceptible materials to be used for work in progress and ensure both new and partially completed work is protected against frost.

When necessary the use of heaters should be considered to prevent aggregates and other materials from becoming frozen and to prevent frost occurring in completed work. Note however that excessive application of heat can give rise to rapid drying and the risk of shrinkage or cracking occurring.

PRECAUTIONS TO TAKE TO AVOID DAMAGE TO CONCRETE WHEN THE TEMPERATURE IS NEAR FREEZING

For foundation and oversite concrete, either,

◆ Do not concrete foundations or oversite when the ground or hardcore is affected by snow, ice or frost, or,

◆ If work has to proceed during long periods of cold weather, precautions shall be taken to keep the ground and new foundation work above freezing.

If the ground or oversite is frozen, STOP WORK. Work built on frozen ground can be severely damaged by movement when water in the ground thaws.

Consider erecting a cover over the whole area of work. Provide heat to create a frost free environment.

Site mixing

If the temperature reaches 2°C (36°F) and is falling, concrete work should only proceed if:

◆ The aggregate temperature is above 2°C (36°C) and is free from frost or snow, and,

◆ Water for mixing is heated, but not in excess of 60°C (140°F), and,

◆ The cast concrete is properly protected.

Stop work if the aggregates are frozen unless they can be thawed. Remember that covering aggregate with tarpaulins will not stop severe frost from penetrating into the aggregate. If work is to continue it may be necessary to steam heat aggregate or use hot air blowers below covers.

Do not rely on heated mixing water to thaw frozen aggregates. The amount of water in a mix is only a small proportion of the total volume of the mix. Very cold aggregate can absorb heat from water while still remaining frozen.

STOP WORK ON ANYTHING WHICH WILL BE AFFECTED BY FROST WHEN THE TEMPERATURE IS AT 2°C (36°F) AND EXPECTED TO FALL TO FREEZING. CHECK THE TEMPERATURE THROUGHOUT THE DAY.

Ready mix concrete

The temperature of the concrete when delivered must be at least 5°C (41°).

The cast concrete must be properly protected. As a guide, 50mm of insulation held down firmly at edges will give protection to oversite concrete from slight overnight frost. If very severe frosts are expected, insulation by itself would not be adequate and it would be necessary to use heaters below covers.

Precautions to take to avoid damage to masonry

Materials which are affected by snow, ice or frost must not be used.

Do not proceed with any work when air temperature is 2°C (36°F) and falling.

Continue to work only if heat is provided under cover to keep the masonry above freezing.

SNAGGING

CARRY OUT CHECKS TO IDENTIFY DEFECTS AND UNCOMPLETED WORK

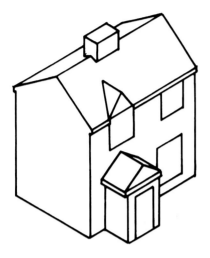

HAVE SNAGGING COMPLETED PRIOR TO PURCHASER'S ARRIVAL.

Identification of defects should be carried out at all stages of construction.

General snagging procedure
- snag each property before handover in a systematic sequence so as to identify any defects (see below)
- write down the defects
- pass a list of remedial work to the responsible trades foreman or subcontractor with a suitable written deadline for completion of the work
- check that remedial work has been completed correctly
- check that all items listed in 'the Schedule' have been provided.

Snagging involves looking at all parts of the property in a systematic way. Establish a snagging system and use it consistently.

Example
- check each external elevation in turn - like a book - from top to bottom and left to right
- check each room in a consistent way - either clockwise or anti-clockwise.

This document may be used as a checklist - all appropriate checks should be carried out together with any others relevant to the specific dwelling.

Snagging should be undertaken for the following:

A Roofs

B External walls

C Attic spaces

D All interior spaces

E Kitchens, bathrooms, etc.

F Connection and testing of services

G Remedial measures

A ROOFS

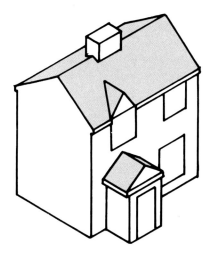

Guidance.
Check that:

Pitched roof coverings
- roof tiles or slates are complete
- bedding mortar at verges, ridges, hips and valleys is complete and free from cracks
- proprietary venting tiles are present, where needed
- proprietary dry verge and ridge systems are complete
- there are no cracked or slipped tiles or slates
- the roof covering is free from mortar
- the line is even and the gauge is correct
- the covering is nailed/clipped/ fixed, as appropriate.

Flat roof coverings
- chippings are distributed evenly
- reflective coating
- roof finish is free from blisters and cracks
- roof is free from ponding
- upstands and flashings are in correct locations and of sufficient height
- there are falls to outlets, eaves, gutters, etc..

Dormer roof and cheeks
- tiling is complete and free from cracked or slipping tiles
- flashings and soakers are correctly installed
- flashings and edge details are complete.

Chimney stacks
- flashings are correctly installed, wedged and pointed up
- pointing and haunching is complete
- cappings are bedded or sealed correctly
- throatings to cappings are formed correctly
- metal tray D.P.C. installed in brickwork chimney.

Flues
- flues are installed correctly
- cappings are sealed to flues correctly
- terminals are installed correctly.

Abutments with walls
- flashings are complete and fixed firmly
- flashings have sufficient lap
- flashings are wedged and pointed up
- weep holes to stepped trays are not blocked
- stepped trays are provided as necessary.

Eaves
- ventilation openings or proprietary ventilation units are adequate
- fly screens are fitted over large ventilation openings
- soffits, etc. have been painted or stained.

Rainwater drainage
- valley gutters are clean and free from debris
- gutters, brackets and stop ends are positioned properly, complete and undamaged
- falls to outlets are correct
- roof underlay laps into gutter
- joints in gutters are installed correctly and not leaking
- debris has been removed.

B EXTERNAL WALLS

Checks for external walls

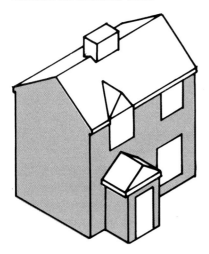

Guidance

Check that:

Brickwork
◆ pointing is complete
◆ brickwork is clean and free from mortar splashes
◆ mortar is of a consistent colour.

Rendering
◆ junctions with half-timber neo-Tudor designs are correct
◆ details at openings are acceptable
◆ bell castings and beads are finished correctly
◆ texture and finish are even and consistent.

Claddings
◆ details are correct and complete:
  - at openings
  - at external corners
  - abutting masonry
  - at terminations, e.g., top.

Gutter downpipes
◆ fixings and joints are correct
◆ components are complete and undamaged
◆ downpipes are connected to outfall, where required.

Waste pipes
◆ making good has been completed neatly where waste pipes pass through the wall.

Overflow pipes
◆ overflows can be readily seen
◆ pipes are not too short at low level
◆ making good is complete around pipes
◆ overflow from unvented hot water system has been installed correctly to provide safe discharge.

Balanced flue terminals
◆ guards have been installed correctly
◆ terminals are located away from fascia, soffit, windows and doors in accordance with boiler manufacturer's installation instructions.

Windows and doors
◆ window frames are free from chips and splits
◆ cills and drip moulds are undamaged
◆ doors are undamaged
◆ mastic is present and applied properly, where specified
◆ glazing method is correct
◆ glass is not cracked or scratched
◆ glass is reasonably clean and free from paint splashes
◆ putty is complete
◆ beads are fixed securely
◆ safety glass is used in risk areas.

Paintwork and staining
◆ work is complete, especially at high level and underneath cills
◆ top coat has been applied to painted surfaces.

C ATTIC SPACES

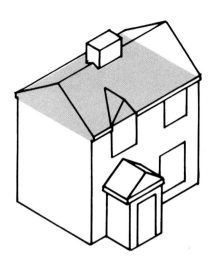

Guidance
Check that:

Masonry wall/roof junctions
◆ ends of purlins, binders, etc. are properly supported and built into masonry where required
◆ making good is complete.

Separating wall/roof junction
◆ brickwork or blockwork is made good up to roof line
◆ mortar joints are fully filled with mortar
◆ firestopping to underside of underfelt/sarking is complete and leaves no gaps
◆ fire-stopping between top of sarking/underfelt and bottom of roof covering has been provided.

Underfelt/sarking
◆ underfelt is complete and undamaged
◆ underfelt is flashed properly around openings, soil stacks, ridge vent units, etc..

Trussed rafter roofs
◆ bracing and restraint strapping is complete and fixed adequately
◆ support is adequate to water tanks
◆ tags are visible.

Roof ventilation
◆ eaves ventilators are fitted where needed
◆ ventilation path is not blocked by insulation.

Insulation
◆ insulation is complete, and to the required thickness.

Tank and pipe insulation
◆ insulation to tank is complete, including lid
◆ there is no insulation below tank
◆ water pipes and overflows above loft insulation are lagged fully, and adequately supported.

Access to roof space
◆ access hatch is insulated, is the correct size and fits the opening correctly

Flue pipes (where present)
◆ flue pipe connections are correct (should be socket uppermost)
◆ flue pipe supports are directly below each socket
◆ combustible materials are a suitable distance from flue pipes
◆ flue pipes are connected correctly to ridge terminal.

Vent pipes
◆ vent pipes are supported adequately through roof space and connected to a terminal
◆ insulation to extract ducting is fitted and ducting is terminated properly to the outside air.

Roof spaces
◆ debris and rubbish has been removed.

D ALL INTERIOR SPACES

Check inside all rooms, staircases and circulation area.

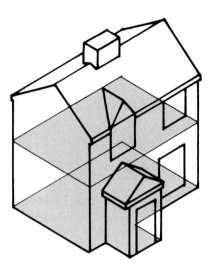

Guidance
Check that:

Ceilings (and walls which are dry lined)
◆ cornices and corners are even, true and made neatly
◆ paintwork or decorative finish is acceptable and free from blemishes
◆ there are no damaged areas
◆ joints do not show through.

Walls - Plasterwork
◆ the surface of the plaster work is to an acceptable standard
◆ plaster has been completed neatly around pipes, light switches, socket outlets, etc.
◆ paintwork is to an acceptable standard
◆ junctions, corners, reveals and margins are all even and true.

Windows
◆ paintwork or stain is complete, including exposed underside of window board
◆ frames and window boards are undamaged
◆ glazing is free from paint splashes, not cracked or scratched
◆ windows open and shut properly
◆ trickle ventilators are in working order
◆ window opening lights are not warped or twisted.

Doors and frames
◆ doors open and shut properly
◆ paintwork is complete, including hidden surfaces
◆ paintwork is free from blemishes
◆ doors are not warped or twisted
◆ there is an even gap between doors and frame
◆ appropriate type of fire door between house and garage.

Ironmongery
◆ ironmongery is complete and in working order
◆ ironmongery is free from paint splashes
◆ spare keys are tagged and identified.

Skirtings, architraves, etc.
◆ woodwork is undamaged and free from defects
◆ paintwork is complete and free from blemishes.

Ventilation
◆ Permanent ventilators are installed in all rooms as required.

D **ALL INTERIOR SPACES**

continued.

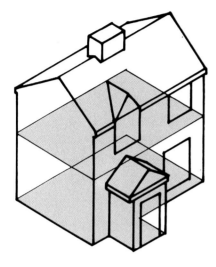

Guidance
Check that:

Floor decking
◆ floor/skirting junction is finished neatly
◆ boarding or decking is fixed adequately
◆ boarding or decking is fitted neatly around pipes/services.

Floor finishes
◆ all floor tiling is complete, especially around cupboards, fittings, doorways
◆ tiling or sheet floor finishes are flat (i.e. no curling edges or bubbles)
◆ there are no paint splashes
◆ screeds are free from rough patches, cracks or hollow areas.

Services
◆ sufficient clips or brackets support pipework
◆ sufficient clearance prevents pipes squeaking, etc.
◆ light switches and socket outlets are fixed adequately and free from paint splashes
◆ the correct number of lighting outlets, switches and socket outlets have been provided
◆ radiators are fixed and finished properly and left undamaged and not leaking

Fireplaces
◆ fire surround is clean and undamaged
◆ debris and rubbish has been removed
◆ the throating is clear
◆ where necessary, there is a sufficient source of air for combustion.

Cupboards
All items above have been checked and in addition:
◆ shelving is complete, including in hot press (where provided)
◆ debris and rubbish has been removed
◆ where a hanging rail is provided, the fixings are adequate.

Staircases
◆ balustrading, newels, handrails are all complete and fixed securely
◆ finish to staircase joinery, etc. is clean and undamaged
◆ staircase steps have even rise and going.

General
◆ debris and rubbish has been removed
◆ floors are clean.

E KITCHENS, BATHROOMS, ETC.

Check inside:
Kitchens, bathrooms, shower rooms, wc's and utility rooms.

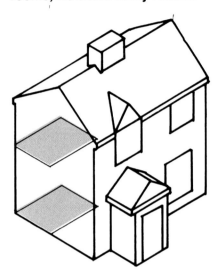

Guidance
Check that:

Sanitary fittings
- sanitaryware is undamaged, working satisfactorily and securely fixed
- labels are removed (except for instructions to purchaser)
- bath panel is fitted
- taps and showers are in working order
- fittings are clean
- W.C. flushes properly
- junctions with walls are watertight.
- Provide plug and chain to sinks and bath.

Kitchen units
- units are not damaged, chipped or stained
- units are fixed adequately, especially wall units
- units are free from paint splashes
- taps are in working order
- units are clear of debris and builders' rubbish
- worktops are not scratched or damaged
- doors and drawers open and close properly
- doors are hung square to units.

Plumbing
- there are no leaks, especially at traps and wastes
- pipes are clipped adequately
- cylinder is insulated, hot water pipes insulated.

Drainage
- where long pipe runs have been used, there is adequate provision for rodding.

Electrical
- the cooker control is working
- ventilation units are fitted correctly and in working order
- ELCB circuits are identified.

Gas
- gas points are installed correctly.

Appliances
- appliances are undamaged
- appliances are in working order
- operating instructions have been provided.

Wall tiling
- tiling is complete and undamaged
- grouting is finished neatly and tiles cleaned
- seal between tiling and fittings is complete
- making good around pipes is complete
- patterns are made correctly
- edging details are complete
- no discoloured tiles have been used.

Other finishes
- grilles are fitted over vent openings.

F CONNECTION AND TESTING OF SERVICES

Checks on the services installation.

Guidance
Check that:

Mains connection
◆ mains are properly connected:
- drainage
- water
- electricity
- gas.

Drainage
◆ drainage tests have been carried out
◆ drains are free of obstructions
◆ drains are connected to the proper outfall
◆ gullies are cleared and filled.

Heating systems
◆ systems are in proper working order
◆ the heating installer has carried out performance test and made any necessary adjustments, for example on:
- operation of the thermostats and programmers
- balancing of radiators
- pump settings
◆ combustion ventilation supply is provided, where required.

Smoke detectors
◆ detectors are installed.

G Remedial measures

Carrying out remedial measures.

Remedial measures are best undertaken before handover.

Where this is not possible, arrangements should be made with the purchasers which are mutually acceptable.

BUILDER TO SNAG BEFORE PURCHASER MOVES IN.

RADON IN BUILDINGS

INTRODUCTION

This guide is substantially derived from 'Radon in Buildings' by kind permission of the Environmental Research Unit, St. Martin's House Waterloo Road, Dublin 4.

This guide is intended to inform designers, householders and other building owners about the radon problem and to help in deciding if there is a need to take any action to reduce radon levels in their homes or other buildings. It explains what radon is, how it enters buildings and what effect it may have on health. Reference is made to some of the usual ways of reducing the level of radon and guidance is given on some sources of assistance. It is not intended as a comprehensive document covering all aspects of radon and indoor radiation exposure, but rather as a source of basic information to assist in making informed decisions on the subject.

Further research into the extent and methods of reducing radon concentrations is continuing, and designers should make themselves aware of the most current recommendations.

Further information may be obtained from a publication by the Building Research Establishment (BRE), Garston, Watford WD27JR, England, titled "Radon Guidance on Protective Measures for New Dwellings" 1991.

(1) FACTS ABOUT RADON

The main sources of radiation are of natural origin. The combined effect of all natural radioactivity contributes considerably more to the overall dose (87% of total) than do all of the artificial sources including medical diagnostics, nuclear energy production and weapons testing. Exposure to any form of radiation may cause harm to the exposed person or his or her descendants, and there is no intrinsic difference between artificial and natural radiation in their biological effects. In contrast to the demand for strict controls over exposure to radiation from artificial sources, relatively little attention has been given to control of exposure to natural sources. However, in recent years there has been a growing concern among those dealing with radiation matters about the long term effects of radiation at all levels and from all sources and the circumstances where exposure could, and probably should, be reduced. This is particularly so in the case of indoor exposure to radon and its decay products which account for over 30% of our total radiation.

Radon is a natural radioactive gas that has no taste, smell or colour and requires special equipment to detect its presence. It is found to some degree in all soils and rocks but the amount can vary in different parts of the country and at different times of the year. Radon is formed in the ground by the radioactive decay of small amounts of radium which itself is a decay product of uranium. Radon has a half-life of 3.8 days, which is the time it takes for one-half of the radioactive atoms to decay to the next member of the decay sequence. Being a gas, it can move through porous media such as fractured rock and soil and some is exhaled at the surface. When this occurs in the outdoor air it is quickly diluted in the atmosphere to low and harmless concentrations. However, once it percolates into an enclosed space, such as a building, it can accumulate to dangerous levels because dispersion is restricted by limited ventilation. The concentration will depend on the radon levels in the underlying soil, the construction details of the building and the available ventilation. It may also be introduced indoors by way of ground water supplied from a well, or from building materials containing traces of radium, but normally the amounts from these sources are not of any consequence in this country.

(2) INDOOR RADON — POINTS OF ENTRY

The levels of radon in indoor air appear to have increased in recent decades, most likely due to the desire for "tighter" building enclosures in order to reduce energy consumption.

These levels depend mainly on the concentration of the sub-floor soil gas and the available entry points in the ground floor area of the building. As these factors usually vary from building to building each case must be considered separately. The more fragmented and porous the underlying rocks and soil the greater the amount of radon gas that can rise to the surface. The gas can enter a building in a convective flow through cracks and holes in the floor area and any gaps around service pipes and cables (figure 1).

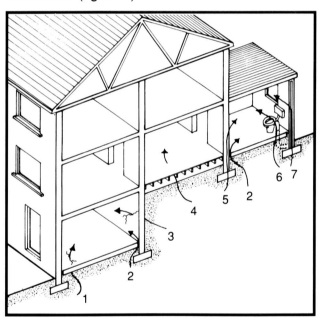

Fig. 1 Major radon entry routes.

1 Cracks in solid floors.
2 Construction joints.
3 Cracks in walls below ground level.
4 Gaps in suspended floors.
5 Cracks in walls.
6 Gaps around service pipes.
7 Cavities in walls.

It is usually pressure-driven due to the slightly lower indoor air pressure compared with that under the floor, a result of wind and temperature difference. As might be expected, elevated levels of radon resulting from soil gas are found mainly in basements and at ground floor levels. Also, radon gas is nine times heavier than air, and therefore tends to remain close to the ground. Radon is not normally a problem in the upper stories of high rise buildings and if an elevated level is found it is likely to have been caused by emission from the building materials used.

(3) OCCURRENCE IN IRELAND

Radioactivity in air resulting from radon is measured in becquerel per cubic metre (Bq/m^3). This unit of measurement indicates that radon is present at a concentration that emits one particle of radiation per second in a cubic metre of air. A survey carried out by Dr. J.P. McLaughlin of University College Dublin of a random sample of approximately 1300 houses in the State shows a median level of indoor radon of about 35 Bq/m^3 throughout the country. However, levels in excess of 400 Bq/m^3 were found in 1.5% of cases with individual peaks rising as high as 1700 Bq/m^3. Most of these were located in parts of counties Clare, Galway, Mayo and Cork, but even in these counties the vast majority of the sample houses had a low radon level. A more recent survey of over 500 houses in some western counties was carried out by the Nuclear Energy Board and U.C.D. to identify the distribution of elevated concentrations in these areas. The results indicate that about 2.8% had radon levels above 400 Bq/m^3 and 9.4% were above 200 Bq/m^3.

(4) REFERENCE LEVEL

Having a screening measurement carried out is the only way of knowing if a building has a radon problem. A recommended Reference Level for Ireland has been set by the Government at 200 Bq/m^3 for the annual average radon gas concentration in an existing house. Above this level action should be taken to reduce it. The level of 200 Bq/m^3 is also intended to apply to all future dwellings which ideally should be constructed so that radon concentrations are as low as reasonably practicable and be at least below this level. In existing dwellings, for levels up to 500Bq/m^3 it would be desirable to take action within a few years, and, where levels are over 1000 Bq/m^3, within a year or so.

(5) RISKS OF LUNG CANCER

The reason for concern about radon is its association with an increased risk of developing lung cancer. There is considerable evidence available both from studies of underground uranium miners and from animal experiments to make this link with certainty. Radon being radioactive disintegrates and gives off decay products known as daughters or progeny which are also radioactive. These are minute particles which when released in the air may be inhaled and deposited in the lungs. As they in turn decay they give a radiation dose to the lung tissues which may eventually cause lung cancer. This is not a matter that should cause immediate anxiety as it takes many years for the disease to develop and is normally considered as a lifetime risk. Not everyone exposed to elevated levels of radon is certain to develop lung cancer and while few people these days spend a lifetime in the same home it would be foolish to ignore the risk completely.

For most people, the risk of developing lung cancer from radon is insignificant compared with other everyday risks. Nevertheless, despite the lack of complete agreement among experts on the precise risk, it has been calculated that exposure to the Reference Level of an annual average of 200 Bq/m^3 corresponds to a lifetime risk of lung cancer of about 2 %. The normal lifetime risk of contracting lung cancer in Ireland from all sources is about 3%.

While smoking has been accepted as the single most important cause of lung cancer, international estimates now suggest that between 5% and 10% of lung cancer deaths may be caused by indoor radon exposure alone. There is also strong evidence which indicates a much higher risk from radon for cigarette smokers than non-smokers. The National Radiological Protection Board in the U.K. in a recent publication puts this risk at 10 times that for smokers than for non-smokers at all levels of exposure. This arises from a synergistic or multiplicative interaction of both carcinogens, which means that the combined effect exceeds the sum of two effects taken independently.

(6) DOSE

Exposure to a given concentration of radon gas may be converted into an annual radiation dose. The unit used to measure this dose is the milliSievert (mSv). The average annual dose to an Irish person from naturally occurring radiation is about 2.5 mSv. Exposure to the Reference Level of 200 Bq/m^3 over a full year would

correspond to an annual dose of 5 mSv. The radiation limit currently enforced by the Nuclear Energy Board in its general licensing to protect the public is 1 mSv. For comparison, the dose arising during the first year from the Chernobyl accident in 1986 was between 0.1 and 0.16 mSv in Ireland and the annual dose to the average consumer of fish from the Irish Sea is 0.002 mSv.

(7) METHODS OF DETECTION

The two most common devices used for measuring indoor radon concentrations are the charcoal canister and the alpha track detector. The charcoal canister is a small container of activated carbon which adsorbs radon. It is exposed in a living area for about a week and then sent to a laboratory for analysis. The alpha track detector gives a more accurate reading of the average exposure but must be left in place for a longer period, usually three months, to cover the widely fluctuating daily and seasonal variations. The detector consists of a small container which allows the alpha particles released by the radon to come in contact inside the container with a small piece of special plastic in which tracks are formed by the radiation striking it. After exposure for the recommended time it is also sent to a laboratory for analysis. The initial screening measurements may indicate that there is no need for further action, but in some cases it may be necessary to take measurements over a longer period to get a more accurate estimate of the average level. Remedial action should be undertaken only after long-term measurements have been taken using alpha track detectors, which is the method recommended by RPII as the shorter period measurements obtained with charcoal canisters may not give a reliable indication of average levels. There are other techniques requiring operation by trained personnel which can be used to give "instant" readings, but they are more expensive and, due to the normal variation in concentrations over a short period of time, would be less reliable in determining the average radon level.

It is possible to take radon measurements in the ground on a prospective building site, but the results may be of limited value. Alfa track detectors may be buried in holes about 600mm deep and recovered after exposure for one week. If high readings are found there is no doubt about the need for preventative measures. Low readings in the ground, however, may not be taken as a guarantee of low concentrations inside a future building on the site, as any excavation necessary during construction may increase the release of soil gas which could result in high indoor concentrations. After construction it would probably be necessary to have indoor measurements taken in both cases to find out the actual level inside the building.

(8) PREDICTION AND MEASUREMENT OF LEVELS

One of the difficulties at present with the occurrence of radon gas is that no reliable method has been found for identifying the geographical areas most at risk. There is evidence linking high radon levels with underlying areas of uranium-bearing granite, shales, phosphate and certain sandstones, but this pattern is not entirely reliable which makes it difficult to prepare maps predicting areas of high concentrations based on geological data. As an alternative, national surveys based on map grids are extremely expensive and lengthy exercises and result in only very general indications due to the variations in radon concentrations that may occur within small areas. In the absence of trustworthy maps or other methods of indicating places most at risk, it rests with individuals to consider having their sites and buildings tested if they are preparing for new building works or are concerned about existing structures.
Measurements may be arranged by applying to either of the following:-

Radiological Protection   Physics Department
Institute of Ireland   University College Dublin
3 Clonskeagh Square   Belfield
Clonskeagh Road   Dublin 4.
Dublin 14.

A measurement will involve receiving by post an alpha track detector with instructions on how it should be exposed in a building for some months before being returned for processing. The results will then be sent by post to the applicant and will remain confidential. The charge for this service is currently £15 (i.e. 1993). The applicant will be told either that no further action is necessary or, where a potential hazard has been indicated, that additional testing over a longer period would be advisable. Should it be necessary to undertake remedial measures to reduce the levels found, it may be

necessary to seek professional advice from an architect, radiation scientist or engineer who is properly qualified to deal with this type of work, in order to decide on the appropriate action.

(9) CORRECTIVE OPTIONS

Compared to existing buildings, it is relatively easy when designing a new building to take precautions to deal with radon should the problem arise. The selection and design of a cost-efficient reduction system will depend on a number of factors specific to the individual building.
Reduction techniques fall into two main categories: those aimed at preventing radon entering the building and those aimed at removing radon or its decay products after entry. Techniques which prevent radon entry include: sealing soil gas routes into the building, sub-floor ventilation to draw or force soil gas away from the building before it can enter, and adjustment of the air pressure inside the building to reduce or reverse the driving force which assists the entry of soil gas. Techniques which remove radon after entry include: ventilation of the building, and air cleaning devices to remove radon decay products. For persons who smoke cigarettes and are concerned about lung cancer attributed to radon, probably the most effective first action would be to give up smoking. For others, it may be prudent to begin by installing the simplest and least expensive method which offers reasonable potential for achieving the desired reduction. If this does not work, the system could then be expanded in a series of pre-designed steps until the level is reduced to that required.

(10) REDUCTION TECHNIQUES

A solution to high indoor concentrations of radon should ideally:

1   Limit the infiltration of soil gas;

2   Avoid if possible "active" solutions which require long term energy expenditure (fans that must run constantly to remove the radon);

3   Not compromise attempts to limit energy consumption in the house.

(A) Sealing

Tests have established that the concentration and pressure of soil gas is the dominant factor contributing to indoor radon problems and that the most common method of infiltration is through cracks and other openings in the ground floor and adjoining walls, particularly when the floor slab is in contact with the ground.

(i) Existing Buildings

To prevent this infiltration in an existing building the normal recommendation is to make the floors and walls a more effective barrier by sealing all the points of entry. Unfortunately, this is easier said than done. Experience has shown that missing even minor openings in the sealing process will compromise the exercise. Since the gas flow is pressure driven, it is similar to mending punctures in a tyre tube - if you miss sealing just one hole you may still have a flat tyre. Nevertheless, by a combination of careful inspection and thorough workmanship, it should be possible to seal all major entry routes and reduce the main inflow of radon and, at worst, end up with the equivalent of a slow puncture rather than a totally flat tyre. In many cases it will not be possible to carry out a full inspection due to existing floor coverings, skirtings etc. and the chances of successful sealing will be further reduced.

Materials for this sealing work should be flexible, permanently elastic and capable of adhering to a variety of surfaces. Buildings are to some degree dynamic in the sense that minor movements may occur from year to year and, unless the sealant is able to flex with the building, cracks will reappear. High quality sealants such as silicone, polysulphide and polyurethane are those most likely to be successful.

In existing buildings with timber floors having wall to wall carpeting it may be possible to provide a barrier by installing a sheet of 1000 gauge polyethylene film directly on the timber flooring under the underlay or foam-backed carpet. Any joints in the film as well as junctions between floor and walls must be taped or otherwise sealed to provide an airtight joint. Precautions must be taken to avoid tears in the film and in general, great care is required in the workmanship. Other existing floors with relatively impervious sheets or tile materials may

only require sealing at joints and edges to provide a barrier. In most cases it will be difficult, if at all possible, to achieve a totally airtight seal, but nevertheless, this must be the objective if radon entry is to be reduced. *(HomeBond do not recommend the covering of timber floors with plastic sheeting, because of the risk of timber decay).*

(ii) New Buildings

In a new building it should be possible to provide a barrier against rising gas by installing an impermeable membrane across the total floor and wall sections. It is important that this barrier prevents radon from rising in wall cavities and in voids of hollow concrete block walls. To achieve this it will be necessary to modify traditional construction practice and ensure a higher than normal level of workmanship. For example, in a cavity wall if the damp proof course is to act as a barrier the normal separate damp proof course in each leaf should be replaced with a single stepped damp proof course across the cavity (Figures 2 and 3) having joints lapped and sealed or with a membrane carried through to the outside face (Figure 4). The membrane and damp proof course arrangements shown in the Figure are possible solutions and illustrate the principle of barrier protection: other arrangements may be equally valid. Precautions should be taken to minimise possible differential movement between walls and concrete slabs and to make provision for this in the membrane at wall and floor joints. Many plastic films such as polyethylene, PVC etc. are suitable as membranes provided they are of adequate thickness and any joints are formed with airtight seals. Where possible, service pipes should not penetrate the radon-proof membrane and where this is not possible the penetrations should be provided with an airtight seal.

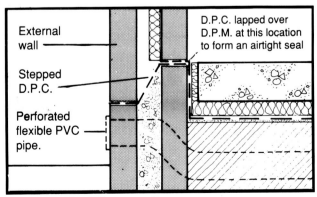

Figure 2: Wall/floor section

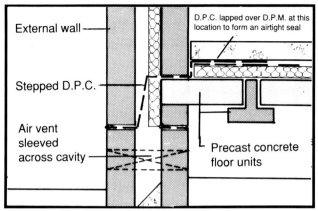

Figure 3; Wall/floor section.

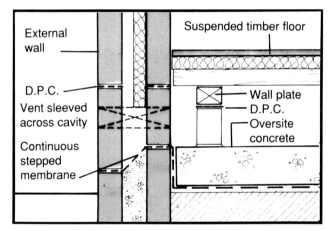

Figure 4: Wall/floor section.

Alternatively, it may be possible to rely on an in-situ concrete ground floor slab on its own to provide a barrier. The Swedish Council for Building Research has indicated that a concrete slab of conventional good quality, provided it is free from cracks and holes, presents sufficient resistance to radon, even where there is a high sub-slab concentration of the gas. In such cases consideration should be given to the use of steel or polypropylene fibres in the concrete mix in order to control the incidence of plastic shrinkage cracking. Suitable airtight seals at slab perimeter to damp proof course etc. would, of course, be necessary.

(b) Depressurization

After sealing or providing a barrier, the most common way of attempting to cope with infiltration of radon is to eliminate the pressure gradient between the soil and the building. In a building with a concrete ground floor slab in direct contact with the soil, this will usually mean

providing, in the permeable hardcore layer under the slab, a sump or collection chamber into which the radon gas is drawn and from which it is piped to the outside air. An electrically operated fan will normally be necessary to provide the suction, although in certain circumstances the "stack effect" of a pipe taken up through the building may be able to reduce the underfloor pressure sufficiently. A fan operated system is far from being an ideal solution as, apart from the cost of installing and running a fan constantly, the noise and maintenance aspect cannot be ignored. However, despite all its drawbacks it is still the most reliable of current methods when radon levels are very high and may be the only practical option in severe cases.

(i) Existing Buildings
In an existing building this remedy can be both troublesome and expensive. It will involve breaking through the existing slab to form a sump, installing the associated piping, reinstating the slab and existing damp proof membrane and sealing the junction between the old and new concrete. As a solution for an existing building, this technique should be considered only when other options give little hope of success.

In an existing building with a very permeable hardcore layer under the floor it may be possible to make a pipe connection to this layer at a point outside of and adjoining an exterior wall. This would involve breaking through the rising wall and inserting a pipe to which a fan would be attached to reduce the underfloor pressure and draw the soil gas out into the open air.

Also, where the ground adjoining an existing building is very permeable it may be possible to form a sump outside of and close to the building and have a suitable fan fitted to draw soil gas from the ground which may reduce the pressure in the underfloor area.

(ii) New Buildings
Installing a sump when constructing a new building is relatively easy and inexpensive and useful precaution even if later found to be unnecessary.

The normal recommended underfloor sump is usually centrally located and consists of a 600 x 600 x 50mm concrete paving slab supported on loose-laid bricks on edge spaced apart around its perimeter (Figure 5 and 6). A 100mm diameter PVC pipe is inserted into the chamber thus formed and is taken through an exterior wall where it is turned up and capped until it is necessary to extend it and fit an extract fan. Alternatively, the exhaust pipe and fan may be located at a suitable place within the building for soil gas extraction to the outside air at a high level. Depending on the size and layout of the building, more than one sump may be necessary.

(c) Ventilation

Ventilation is the process whereby internal air is gradually replaced by outdoor air. Natural ventilation is driven by pressure differences caused by wind or by differences in air density between indoors and outdoors. As a remedy for radon, ventilation should be considered in two ways, - ventilation of the space under the ground floor and ventilation of the internal building space.

(i) Underfloor Ventilation - Existing Buildings
Existing buildings with high radon levels which have a suspended type of ground floor construction are fortunate in having what may be an easily remedied situation. The air space under the floor effectively disconnects it from the ground and offers the opportunity of intercepting the rising soil gas and removing it by ventilation before it can enter the building. The underfloor ventilation normally provided to remove rising ground moisture can usually be increased to allow a sufficient strong undercurrent of outside air to constantly replace the soil gas as it emerges from the ground. As the increased air movement under the floor would have a cooling effect, it may be necessary to install floor insulation to prevent an energy penalty. The ease of adding insulation would depend on the construction details. Should this approach not result in the desired reduction within the building, it would then be possible to resort to "active" depressurization of the sub-floor by sealing all the wall vents and installing an extract fan system to draw air from the space under the floor. *(HomeBond advise that where floor vents are sealed and extract ventilation is provided under timber floors, regard should be had to ensure that the ventilation levels provided are adequate to protect the floor construction against the risk of decay arising from excessive moisture).*

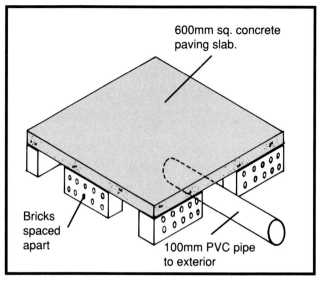

Figure 5: Underfloor sump details.

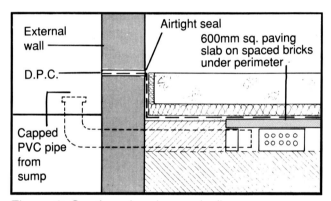

Figure 6: Section showing underfloor sump.

(ii) Underfloor Ventilation — New Buildings
For new buildings, whether there is evidence of high ground radon or not, the use of a suspended type ground floor in either timber or precast concrete makes provision for a remedy if one is needed in the future. Normal underfloor ventilation may be installed initially and, if indoor radon measurements taken after construction show acceptable levels, no further action is necessary. Should the indoor measurements be high, the steps previously described for existing buildings could be taken to the extent necessary. Since the provision of floor insulation would be normal in most new buildings, there would be no need for further insulation work.

In a new building where the preferred form of ground floor construction is a concrete slab on the ground there is an alternative method to the type of under slab depressurization sump previously described. This alternative, while still at experimental stage, would permit a "passive" ventilation approach as a first option and, subject to indoor measurement levels, a second option of using an "active" fan assisted depressurization system if that becomes necessary.

The construction detail would provide for a number of 100mm UPVC perforated land drainage flexible pipes embedded in the hardcore layer immediately below the floor slab. (Figure 2). These pipes would penetrate through, and finish flush with, the outer faces of the external wall at a level just above the finished exterior ground, forming a connecting conduit through which a current of air could flow. Such cross ventilation could, with sufficient air movement in the conduit, draw into it the rising radon gas and transfer it to the outside air. The number and arrangement of pipes necessary would depend on the house exposure to the prevailing winds, the construction details and the general layout. Interconnection of the main pipe runs within the floor plan area would be an advantage. It may be necessary in some cases to adjust the ground levels adjoining the exterior walls and also the hardcore layer to permit the pipe end openings to be located above the exterior ground or paving levels. Should this natural ventilation system not produce the required reduction, the ends of the pipe conduits at the exterior walls could be sealed where necessary and an "active" fan system installed to depressurize the underfloor layer. In this case the perforated pipe would become an elongated sump performing in the same way as the brick and concrete slab collection chamber previously described, but arguably in a more effective shape. It should be easier to install than the normal sump and offer flexibility where more than one sump may be necessary. The principal advantage of this system is that it offers the prospect of a passive ventilation solution which can be converted in a "fall back" situation to a mechanically operated system.

(iii) Ventilation of Internal Space — Existing and New Buildings

Increasing the ventilation of living areas, whenever possible, by opening windows on two or more sides of a building is the simplest method of reducing high radon concentrations. This will not always be possible for security reasons and in cold weather would be impractical due to discomfort and the increase in heating costs. This additional heating cost could be significantly reduced by installing a mechanically balanced supply/exhaust ventilation system with heat recovery. This would introduce extract air at approximately the same rate, resulting in a neutral pressure within the building. To operate successfully, a balanced system requires a reasonably well sealed building in order to control air movement, and for radon reduction, it is essential that there are properly designed and located air inlets and outlets to allow incoming air to mix with room air. Mechanical ventilation of this type may not on its own be able to remedy a high level of radon but, as well as generally improving the quality of indoor air, it could be a useful supplementary tool with other systems. However, these systems are relatively expensive to install and it is unlikely that they would find wide-scale application in this country.

(d) Pressurization — Existing and New Buildings

It is generally accepted that the use of extract fans within a house to improve ventilation is counterproductive in dealing with high levels of radon. The net effect is to reduce the pressure in the house and thereby induce more radon from the soil through any available openings in the ground floor. It should be noted also that the effect of open fires and other combustion appliances is to reduce air pressure within a house and for this reason should ideally be provided with a dedicated source of outdoor air. An alternative approach is to use a fan system to provide a positive pressure throughout the house and in that way reverse the normal inflow of soil gas from under the floor.

In a house with a pitched roof this fan unit could be housed in the attic space and air stream introduced into the house through a diffuser in the ceiling. Each room would be slightly pressurized and air forced out through crevices in windows, doors and other openings, reversing the normal inflow. Compared with outside conditions, the air in the attic would be warmed by solar gain from most of the year during daytime, but during night-time use in very cold weather some pre-warming of the air would be necessary. A temperature controlled small electric heater could remove the chill but would add to the running cost. There is available commercially an attic mounted unit which was designed primarily for condensation control but which could in some cases provide the desired pressurization to control radon entry. This system would likely have most success in a reasonably well sealed house where a constant positive pressure could be maintained with a relatively low powered fan.

Some sources have dismissed this approach on the assumption that warm and moist indoor air would be forced into the building fabric giving rise to rot and condensation staining. This result is unlikely except in an extremely "tight" structure with few openings to allow the air to escape — a situation which is difficult to achieve in practice. For existing houses without other easy alternatives, this system of pressurization may provide a solution and would be less disruptive than many of the alternative remedies.

(e) AIR CLEANING — EXISTING AND NEW BUILDINGS

Another way of removing radon progeny from the indoor air space is by various air treatments. Since radon decay products are solid particles they can be removed by continuously circulating the air through a device which removes them when they are attached to atomized or larger dust particles which are suspended in the air. This method has advantages over other techniques in that equipment for solid particle removal from air is available commercially at low cost and there is the added benefit of the removal at the same time of other air pollutants such as dust and pollen. Several devices are available for removing particulates from indoor air. They can be categorized according to their principles of operation into mechanical filters and electrostatic filters. Mechanised filters collect particles through mechanical forces exerted on them by the air flow and a fibrous filter medium. Electrostatic filters collect particles primarily as a result of electrical forces exerted on them when suspended in an air stream. It is important to note that these systems remove only particles from the air and do not remove any radon gas, which means that the production of progeny through the decay of radon is not affected. While these systems are "active" rather than "passive" the capital and running costs are substantially lower than for advanced mechanical systems.

The main problem with these devices is that, despite their success in removing particles with attached progeny, this does not necessarily reduce the radiation dose to the lungs and may even make matters worse. By removing the particles with the attached decay products, the newly created progeny then find fewer dust particles to adhere to and remain in the more hazardous unattached state. Radon decay products attached to particles deposit upon respiratory surfaces with only a few percent probability, whereas unattached progeny deposit with nearly 100% probability. As the mix of unattached and attached radon progeny is an important consideration in assessing lung dosimetry it is understandable why air cleaning devices on their own are not recommended at present as a solution to a radon problem. They can nevertheless effect some reduction in the radiation dose when used in combination with other more efficient systems and may contribute a positive side effect when filters are installed to remedy some other indoor air problem.

11 CONCLUSION

If elevated levels of radon have been identified or are anticipated and it has been decided to take some action, the first step may be to seek advice. Unless the appropriate measures can be chosen with confidence, professional advice should be sought from an architect, engineer or radiation scientist who is properly qualified to make the decisions.

In existing buildings with a radon problem, the options available to prevent the entry of soil gas are few and tend to be difficult to introduce with a high degree of success. Existing construction details and local conditions will often indicate the most appropriate remedy. Sealing alone may work well where the levels are not very high and the workmanship is thorough, but in many cases it may require a combination of two or more methods, e.g. sealing and ventilation, to achieve the necessary reduction. "Active" methods, such as fan assisted depressurization need only be introduced when other simpler methods have failed and there is an urgency to reduce dangerously high levels.

In new buildings it is relatively easy and inexpensive to take precautions against the problem. Building Regulations require that measures be taken to avoid danger to the health of occupants of new buildings caused by radioactive substances in the ground under buildings. Even in areas where the prospect of trouble from radon would not be considered great it may be a worthwhile exercise to take some preventive action. The provision of membrane barriers would in many cases be merely an extension of the precautions already necessary for the prevention of rising moisture and the addition of a subfloor sump, even if never used, would be cheaply bought insurance.

REFERENCES

Nuclear Energy Board – Information Sheet No. 5 "Radon in Houses".

McLaughlin, J.P. "Indoor Radon: Sources, Health Effects and Control", Technology Ireland, July / August 1990.

Department of the Environment for Northern Ireland "Radon in Dwellings" (1989).

Department of the Environment (UK) "The Householder's Guide to Radon", Second Edition, 1990.

National Radiological Protection Board (UK) – "Board Statement on Radon in Homes" (1990).

U.S. Environmental Protection Agency – "A Citizen's Guide to Radon".

Nazaroff, W. and Nero A – "Radon and its Decay Products in Indoor Air" (1988).

Mueller Associates, Inc. "Handbook of Radon in Buildings" (1988). Prepared for U.S. Department of Energy.

Building Research Establishment – "Radon Guidance on Protective Measures for new Dwellings" (1991).

CONVERSION FACTORS

Conversion factors

Bold type indicates exact conversions. Otherwise four or five significant figures are given.

Quantity Conversion factors

General purpose

Length	1 mile	=	1.609km
	1 chain	=	20.1168m
	1 yard	=	0.9144m
	1 foot	=	0.3048m = 304.8mm
	1 inch	=	25.4mm = 2.54cm

Area	1 square mile	=	2.590km² = 259.0 ha
	1 hectare	=	10000m²
	1 acre	=	4046.9m² = 0.40469ha
	1 square yard	=	0.8361m²
	1 square foot	=	0.09290m² = 929.03cm²
	1 square inch	=	645.2mm² = 6.452cm²

Volume	1 cubic yard	=	0.7646m³
	1 litre	=	1dm³ = 1000cm³
	1 millilitre	=	1cm³ = 1000mm³
	1 cubic foot	=	0.02832m³ = 28.32 litre
	1 petrograde standard	=	4.672m³
	1 cubic inch	=	16387mm³ =16.387cm³
		=	16.387ml = 0.016387 litre

Capacity	1 UK gallon	=	4.546 litre
	1 UK quart	=	1.137 litre
	1 UK pint	=	0.5683 litre
	1 UK fluid ounce	=	28.413cm³
	1 US barrel (for petroleum)	=	159.0 litre
	1 US gallon	=	3.785 litre
	1 US liquid quart	=	0.9464 litre
	1 US dry quart	=	1.101 litre
	1 US liquid pint	=	0.4732 litre
	1 US dry pint	=	0.5506 litre
	1 US liquid ounce	=	29.574cm³

Mass	1 UK ton	=	1.016 tonne = 1016.05kg
	1 US (or short) ton	=	0.9072 tonne = 907.2kg
	1 kip (1000lb)	=	453.59kg
	1 UK hundredweight	=	50.80kg
	1 short (US) hundred-weight	=	100lb = 45.36kg
	1 pound	=	0.4536kg
	1 ounce avoirdupois	=	28.35g
	1 ounce troy	=	31.10g

Volume rate of flow	1 cubic foot per minute	=	0.4719 litres/s
		=	471.9 cm³/s
		=	0.000 4719m³/s
	1 cusec (cu ft per sec)	=	0.02832m³/s ('cumec')
	1 cu ft per thousand acres	=	0.06997 litres/ha
		=	0.006997m³/km²= 6997cm³/km²
	1 cubic inch per second	=	16.39ml/s
	1 gallon per year	=	4546cm³/a* = 0.004546m³/a
	1 gallon per day	=	4546cm³/d
	1 litre/s	=	86.4m³/d
	1 million gallons per day	=	0.05262m³/s
	1 gallon per person per day	=	4.546 litre/ (person day
	1 gallon per sq yd per day	=	0.005437m³/(m².d)
		=	0.000062928mm/s
	1 gallon per cu yd per day	=	0.005946m³/(m³.d)
	1 gallon per hour	=	4.5461 litre/h
	1 gallon per minute	=	0.07577 litre/s
	1 gallon per second	=	4.5461 litre/s

Heating Temperature

x° Fahrenheit	=	⁵/₉ (x-32)° Celsius

$x° \text{ Fahrenheit} = \frac{5}{9}(x-32)° \text{ Celsius}$

Temperature interval

1° F	=	0.5556K = 0.5556 °C

Energy (heat)	1 British thermal unit	=	1055J = 1.055kJ
	1 Therm	=	105.5MJ
	1 calorie	=	4.1868J
	1 kilowatt-hour	=	3.6MJ
	1 foot pound-force	=	1.356J
	1 kilogram force-metre	=	9.806 65J

Power (also heat flow rate)	1 Btu per hour	=	0.293 07W
	1 horsepower	=	745.70W
	1 ft-lbf per second	=	1.356W
	1 kgf-metre per second	=	9.806 65W
	1 calorie per second	=	4.1868W
	1 kilocalorie per hour	=	1.163W
	1 metric horsepower	=	735.5W

Density of heat flow rate	1 Btu per square foot hour	=	3.155W/m²

Thermal conductivity k value	1 Btu inch per square foot hour degree Fahrenheit	=	0.1442W/(m.K)

Thermal transmittance or coefficent of heat transfer or thermal conductance or U value	1 Btu per square foot hour degree Fahrenheit = 5.678W/(m²K)	

Force per unit area of Stress or Pressure	1 lbf per square foot	=	47.88N/m² = 47.88 Pa
		=	0.04788kN/m²
	1 lbf per square inch	=	6.895kN/m² = 6.895 kPa
	1 tonf per square foot	=	107.3kN/m² = 107.3kPa
	1 tonf per square inch	=	15.44MN/m² = 15.44N/mm² = 15.44 MPa
	1 kgf per square metre	=	9.807N/m² = 9.807Pa
	1 kgf per sq centimetre	=	98.07 kN/m² = 98.07kPa
	1 bar	=	100kN/m² = 100kPa
	1 millibar	=	100N/m² = 100Pa
	1 standard atmosphere	=	101.325kPa
	1 inch of mercury	=	3.386kPa
	1 foot of water	=	2.989kPa

*a (for annum) is the symbol for year.

VENTILATION AND CONDENSATION

INTRODUCTION

Ventilation of a dwelling is necessary to provide an adequate supply of fresh air for persons using the dwelling, to remove water vapour and prevent harmful condensation. The need for ventilation is a consequence of higher modern standards of insulation, combined with an increase in activities generating water vapour within the building. There has also been a reduction in natural ventilation within buildings due to improved draught proofing.

In order to prevent the harmful consequences of condensation it is necessary to ventilate the building adequately to remove the air bearing the water vapour.

Ventilation should be provided in accordance with the guidance given on the following pages. See also the detailed guidance on roof space ventilation given on pages 167 – 172.

The ventilation system should:

◆ Provide an adequate supply of fresh air for persons using the area;

◆ Achieve occasional rapid ventilation for the dilution of pollutants and of moisture likely to produce condensation in habitable rooms containing sanitary appliances;

◆ Extract moisture from areas where it is produced in significant quantities (e.g. kitchen and bathroom).

A ventilation opening can include any means of ventilation (whether it is permanent or closeable) which opens directly to external air, such as the openable parts of a window, a louvre, an airbrick, a progressively openable ventilator, or a wall ventilator but does not include a flue to an open fire. It may include any door which opens directly to external air. Ventilation openings should have a smallest dimension of at least 8mm other than in a screen, fascia, baffle etc., so as to minimise resistance to the flow of air.

Ventilation to domestic dwellings is required in the following locations;

Habitable rooms.
i.e. any room used for living or sleeping purposes but a kitchen having a floor area of less than 6.5sq. meters is treated differently. (See page 254 for heights of habitable room).

Kitchens (less than 6.5m^2).

Bathrooms.

Sanitary accommodation.
i.e. a space containing a water closet.

VENTILATION REQUIREMENTS FOR A TYPICAL HOUSE.

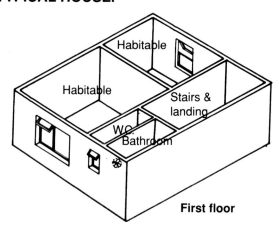

First floor

Sanitary accommodation
Provision should be made for rapid ventilation, either:

(a) By one or more ventilation openings having a total area of at least 1/20th of the floor area of the room, e.g. an opening window, and with some part of the ventilation opening at least 1.75m above the floor level, or

(b) by mechanical extract ventilation capable of extracting at a rate not less than 3 air-changes per hour, which may be operated intermittently, with 15 minutes overrun.

Note: the distinction between bathrooms and sanitary accommodation. In normal housing design, "sanitary accommodation" is any space containing a W.C., and a "bathroom" is any room containing sanitary appliances other than a W.C. If a room contains a bath or shower and a W.C. then the ventilation must be as for sanitary accommodation.

Habitable rooms
i.e. living rooms, dining rooms, bedrooms and kitchens (only if the floor area is greater than 6.5m^2).

Provision should be made for:
(a) background ventilation by a secure permanent opening(s) having a total area not less than 6500mm^2, e.g. a 225 x 225 standard wall vent, located so as to avoid undue draughts or, permavents in windows, and

(b) rapid ventilation by one or more ventilation opening(s) having a total area of at least 1/20th of the floor area of the room, e.g. an opening window(s) with some part of the ventilation opening at least 1.75m above the floor level.

Bathrooms
Provisions should be made for rapid ventilation:

(a) by ventilation opening(s) having a total area of at least 1/20th of the floor area of the room, e.g. an opening window(s) with some part of the ventilation opening at least 1.75m above the floor level, or

(b) by mechanical extract ventilation capable of extracting at a rate of 15 litres per second, which may be operated intermittently.

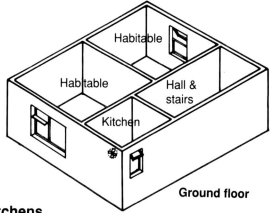

Ground floor

Kitchens
with floor area less than 6.5m^2 (if 6.5m^2 or greater, ventilation as per habitable rooms.

Provision should be made for:

(a) Background ventilation, either:
   (i) by secure permanent ventilation opening(s) having a total area not less than 6500mm^2, e.g. a 225 x 225 standard vent located so as to avoid undue draughts, or permavents in windows, or

   (ii) by mechanical extract ventilation capable of operating continuously at nominally one air-change per hour; and

(b) rapid ventilation, either:
   (i) by ventilation opening(s) having a total area of at least 1/10th of the floor area of the kitchen, e.g. an opening window(s) with some part of the ventilation opening at least 1.75m above the floor level, or

   (ii) by mechanical extract ventilation capable of extracting at a rate of 60 litres per second (or incorporated within a cooker hood and capable of extracting at a rate of 30 litres per second), which may be operated intermittently for instance during cooking.

HEIGHT OF ROOMS

Every habitable room, kitchen and bathroom should be not less than 2.4m in height and the height of such measured beneath any beam in that room and in any bay window shall not be less than 2.1m. Where such a room is immediately below the roof (e.g. dormer construction) its height should be not less than 2.4m over an area equal to not less than one half of the area of the room measured on a plane 1.5m above the finished floor level. See sketch opposite.

Ventilation note:

The ventilation requirements of the Building Regulations as specified in Technical Guidance Document F and also in this section must be adhered to in the construction of dwellint~s.

HomeBond also advise that if new windows, whether single or double glazed, are being installed in existing houses it is vital that there is sufficient ventilation for the rooms, particularly those with fireplaces, to ensure that the fire, whether an open or closed appliance, has sufficient air to draw and also that there will not be a build up of the products of combustion which could be a health hazard.

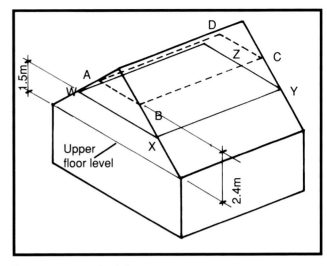

Dormer room height.

Thus: Area of A B C D to be at least half the area of W X Y Z.

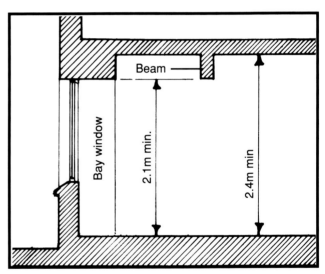

Floor to ceiling height in habitable rooms.

VENTILATION OF HABITABLE ROOMS THROUGH OTHER ROOMS AND SPACES

Two habitable rooms may be treated as a single room for ventilation purposes if there is an area of permanent opening between them equal to at least 1/20th of the combined floor areas.

A habitable room may be ventilated through an adjoining space if -

(a) the adjoining space is a conservatory or similar space, and

(b) there is an opening (which may be closeable) between the room and the space, with an area not less than 1/20th of the combined floor area of the room and space, e.g. an opening window, and with some part of the ventilation opening at least 1.75m above the floor level, and

(c) the space has one or more ventilation openings with a total area not less than 1/20th of the combined floor area of the room and space, e.g. an opening window, and with some part of the ventilation opening at least 1.75m above the floor level, and

(d) for background ventilation there are permanent ventilation openings to the space and openings between room and space, each having a total area not less than 6500mm², located so as to avoid undue draughts, and

(e) the space is not connected to another room.

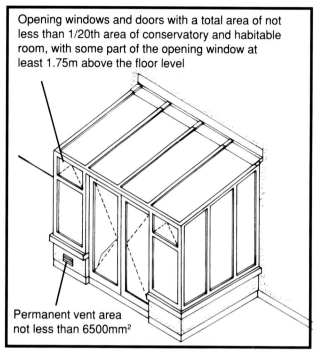

Opening windows and doors with a total area of not less than 1/20th area of conservatory and habitable room, with some part of the opening window at least 1.75m above the floor level

Permanent vent area not less than 6500mm²

Conservatory

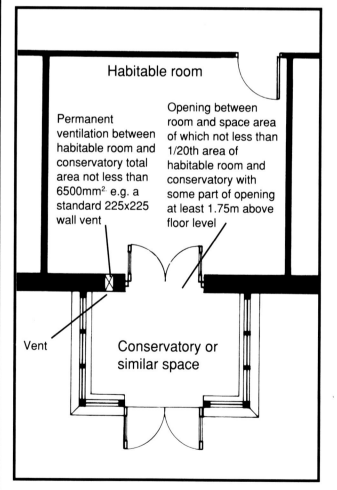

Habitable room

Permanent ventilation between habitable room and conservatory total area not less than 6500mm². e.g. a standard 225x225 wall vent

Opening between room and space area of which not less than 1/20th area of habitable room and conservatory with some part of opening at least 1.75m above floor level

Vent

Conservatory or similar space

Note: If the ventilation opening is to a court then certain restrictions apply and reference should be made to Technical Guidance Document F "Ventilation", of the Building Regulations.

CONDENSATION : INTRODUCTION

Condensation occurs when damp air comes in contact with a cold surface, as warm air can hold more water vapour than cold air.

Moisture produced in kitchens and bathrooms (high pressure areas) will find its way into the lower pressure areas such as bedrooms, where it may condense on the colder wall surfaces as indicated in the sketch.

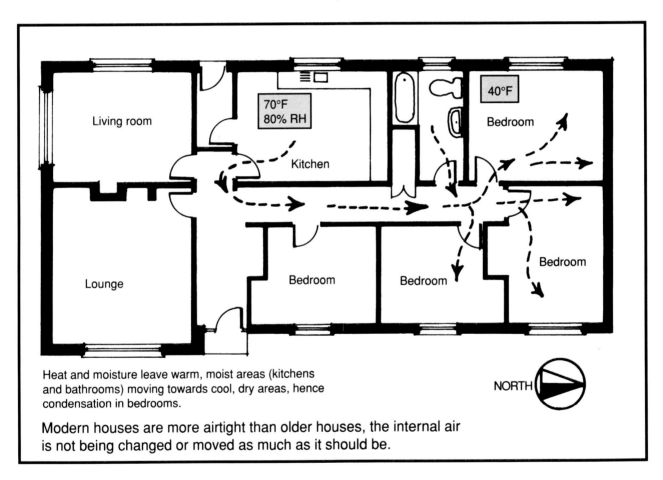

Heat and moisture leave warm, moist areas (kitchens and bathrooms) moving towards cool, dry areas, hence condensation in bedrooms.

Modern houses are more airtight than older houses, the internal air is not being changed or moved as much as it should be.

Stagnant air encourages condensation. Stagnant air pockets are very often the first places where condensation will occur. This can be seen in corners, behind furniture, pictures and in cupboards particularly when placed against external walls. The North wall of a house is often the coldest and most prone to condensation. In rooms with severe condensation it is not uncommon to have to mop up water from the window cill and the jambs every morning due to condensation (condensation channels and weepholes should be provided to all aluminium and timber windows). It is estimated that unheated bedrooms are the cause of at least half of the complaints about condensation.

Sometimes the mould growth can be most severe at the base of a wall, often being confused with rising damp.

Condensation in old houses:
Older houses tended to have more ventilation, larger rooms and higher ceilings, badly fitting doors, windows and floor boards as well as more fireplaces which resulted in better ventilation. Windows are now draught proofed, doors standardised factory made incorporating draught excluders and better fitting. Floors are now concrete or timber covered with carpets or vinyl covering eliminating draughts.

MOULD GROWTH

Once a serious ongoing dampness problem due to condensation has been established in the dwelling, mould fungi (mildew) a parasite freely available in the air as is pollen and yeast etc. can exist and develop on walls, furnishings, bedding clothes, leather goods, shoes, handbags, golf clubs, etc..

Mould first appears as spots or small patches which may spread to form a furry layer usually grey-green, black or brown in colour. There are many types of mould growth, but all grow and spread under similar conditions and all require similar remedial treatment. Three conditions are necessary for mould growth to occur. These are:

1 a source of infection:
   The spores are freely available in the air and can be found in any building.

2 a source of nourishment:
   Dirt, dust, grease, or some organic material is needed but the amount is very small, and consequently almost all surfaces will sustain growth.

3 a damp environment:
   Mould growth requires an atmospheric humidity level in excess of 85% relative humidity sustained for periods up to 12 hours per day. This will result in very serious and widespread growth. At lower levels its growth is slower and it ceases below 70% relative humidity (RM). Mould growth is more marked on North facing walls and in areas where stagnant air pockets exist. Since conditions 1 and 2 above cannot easily be avoided, the deciding factor is relative humidity.

Cleaning of the mould using a suitable toxic wash is required to remove it prior to redecorating, but it must be stressed that such removal is merely cosmetic and not to be recommended as a cure to the overall problem. Cleaning and sterilisation of the materials on, or in which it has occurred is necessary to allow redecoration to continue. It is not always possible to decide whether mould is growing only upon the surface, in a top coat of paint or if it has penetrated further. Where decorations can be stripped off it is usually best to do so.

CONDENSATION

(The following guidance is reproduced by kind permission of the Environmental Research Unit, St., Martins House, Waterloo Road, Dublin 4.)

Condensation is probably the main cause of dampness and mould growth in dwellings. As a result of continuing condensation, walls, ceilings and sometimes floors become damp, discoloured and unpleasant due to mould growing on their surfaces. The following notes explain how condensation occurs and what householders can do to prevent or cure serious outbreaks of it in their homes.

Why condensation occurs.

Condensation occurs when warm moist air meets a cold surface. The likelihood of condensation therefore depends on how moist the air is and how cold the surfaces of the room are. The moistness of the air and coldness of the surfaces depend on a range of factors, many of which are determined by the way the house is used.

When condensation occurs.

Condensation usually occurs in winter. This is because the building surfaces are cold, more moisture is generated within the house and, because windows are opened less, the moist air cannot escape.

Where condensation occurs.

Condensation, which you can see, occurs for short periods in bathrooms and kitchens because of the steamy atmosphere. It also occurs for long periods in unheated bathrooms and sometimes in wardrobes, cupboards or corners of rooms where ventilation and air movement may be restricted. Condensation can occur on materials which are out of sight, for example in roofs.

Condensation forming at junction of walls and ceiling.

What is Important.

In order to prevent or cure condensation problems the following four precautions are important.

1 Minimise moisture production within the dwelling and confine it as far as possible to specific areas, e.g. kitchen, bathroom, scullery.

2 Prevent moist air spreading to other rooms from the kitchen, bathroom or scullery or from where clothes are dried.

3 Provide some ventilation to all rooms so that moist air can escape.

4 Provide some level of heating.

Minimise moisture production

(a) Dry clothes externally when possible.

(b) If using a clothes dryer, provide venting to the outside.

(c) Limit the use of movable gas or paraffin heaters as these types of heaters release large amounts of water vapour into the air and greatly increase the risk of condensation.

(d) Reduce cooking steam as far as possible, e.g. keep lids on saucepans, do not leave kettles etc. boiling for long periods.

Prevent spread of moist air

(a) Good ventilation of kitchens is essential when cooking or while washing clothes. If you have an extract fan in your kitchen, use it when cooking, washing clothes and particularly if the windows mist up.

(b) If you do not have an extract fan, open the kitchen windows and keep the doors between the kitchen and the rest of the house closed as much as possible.

(c) After taking a bath, keep the bathroom window open and the bathroom door shut until the bathroom dries.

(d) Do not use unventilated cupboards for drying clothes.

(e) If you do dry clothes in the bathroom or kitchen, run the extract fan if you have one. Do not leave the door open or the moist air will spread to other parts of the house.

(f) If you have to use a movable gas or paraffin heater make sure the room that the heater is in is well ventilated and sealed off from the rest of the house.

Provide some ventilation.

The easiest method of reducing the moisture content of room air is to provide some ventilation. Ventilation removes the stale moist air and replaces it with fresh air which contains less moisture.

(a) In older houses a lot of ventilation occurs through fireplaces and draughty windows. However, in many modern houses and flats sufficient ventilation does not occur unless a window or ventilator is open for a reasonable time each day and for nearly all the time the room is in use. Too much ventilation in cold weather is uncomfortable and wastes heat. All that is needed is a slightly open window or ventilator. Where you have a choice, open the top part of the window about 10mm.
If more than two people sleep in a bedroom the window should be opened wider, particularly during the night.

(b) If condensation occurs in a room where you have a heater connected to the chimney, you should have the installation checked as the chimney may have become blocked.

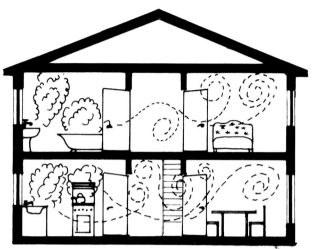

No heating or ventilation accommodates the spread of moist air.

Provide some heating.

(a) Try to make sure that all rooms are at least partially heated. Condensation most often occurs in unheated bedrooms. If you have to leave a room unheated you should keep the window open slightly and the door shut.

(b) Heating helps prevent condensation by warming the room surfaces. It takes a long time for the cold room surface to warm up. It is better to provide a small amount of heating for long periods than to provide a lot of heat for a short period. Houses and flats left unoccupied and unheated during the day become very cold. Whenever possible, try to provide a constant low level of heating all the time.

(c) In houses, the rooms above a heated living room benefit from the heat rising through the floor. In bungalows and some flats this does not happen. Some rooms are especially cold because they have large areas of outside walls and windows or because they lose heat through the roof as well as the walls. Such rooms are most likely to have condensation. Some heating is necessary in these rooms. Condensation is likely if the rooms are not kept above 10°C. When living rooms are in use they should be heated to 20°C if possible.

(d) Insulation reduces the rate of heat loss and helps raise the temperature. However, even in a well insulated house, some heating may be necessary in cold rooms with no indirect heat input.

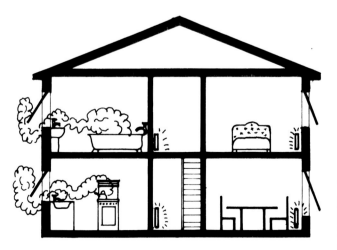

Prevent spread of moist air. Provide ventilation and some heating.

Mould growth

If small black spots appear on the walls or other room surfaces, this is the start of mould growth. Any sign of mould growth indicates the presence of moisture. If the moisture is caused by condensation it is a sign that the level of moisture in the room needs to be reduced or that heating, ventilation or structural insulation, or all three of them, need to be improved.

The mould growth spots should be washed off and the affected area sterilised using a fungicide containing bleach. Make sure that you then tackle the cause of the condensation as recommended under "What is important".

New buildings

New buildings can take a long time to dry out and during the first winter more heating and ventilation will be necessary than in subsequent winters. Excessive temperatures should be avoided to prevent warping of new joinery. With certain types of concrete roofs final drying may only be able to take place inwards. So, do not use waterproof decorations (such as vinyl papers) on the ceiling unless you have been given expert advice that this would not matter.

Effect of extract ventilation on fuel burning appliances.

If you propose to fit an extract fan or otherwise change the ventilation of a room which has a gas or solid fuel appliance connected to the chimney, you should obtain advice from the installer of the heating appliance. This is because there may be a risk of drawing toxic fumes back from the appliance into the room.

EFFLORESCENCE, SULPHATE ATTACK AND FROST ATTACK IN CLAY BRICKWORK

EFFLORESCENCE EXPLAINED

Efflorescence consists of deposits of soluble salts formed on the surface of new brickwork. Although efflorescence is unsightly it is often harmless and seldom persists unless water is able to percolate into the brickwork.

All bricks contain a small percentage of soluble salts. These salts can be harmless unless they are present in bricks which are subject to very damp weather conditions. The problem arises when the bricks become saturated with water. As the brickwork dries out, the soluble salts are carried to the brick surface where they crystallise and are deposited as white crystals. This manifests itself as white patches on the brickwork.

The deposit formed is usually composed of the sulphate of sodium, potassium, magnesium and calcium - not all of these salts being present in any one instance - and sodium carbonate can also appear, usually being derived from the cement in the mortar. Sulphates in bricks may be derived from the clay from which they are made and also from the manufacturing process.

When water is added to the cement the gypsum and alkali in the cement react to form small amounts of sulphates. In a concrete mix where all the water stays in the mix until it is set, all the sulphate will, within a few hours, be fixed in an insoluble form by combination with other constituents of the cement and so will be incapable of contributing to the efflorescence. If, however, the mix is a mortar mix and is placed between dry porous bricks, some of the water will be soaked into the bricks. The sulphate it contains will then be fixed in combined form and will be available to form efflorescence. The characteristic form in which this type of efflorescence appears is as a ring all round the edges of the bricks. Normally, however, the mortar makes only a minor contribution to efflorescence.

Efflorescence on clay brickwork

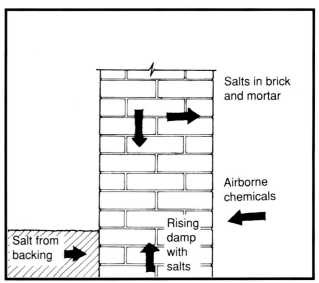

Soluble salt movement

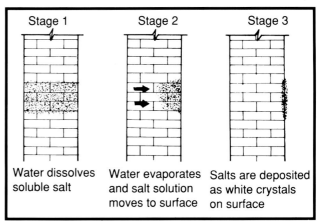

State of efflorescence

WHEN DOES EFFLORESCENCE OCCUR?

Efflorescence is generally a temporary Springtime occurrence which appears as new brickwork dries out for the first time. It is sometimes renewed in the second Spring of a building's life, but this is usually less marked than the first outbreak.

Precautions to avoid efflorescence.

Several precautions can be taken to avoid efflorescence; the main requirement is to avoid moisture getting in to the brickwork during the building stages.

◆ Store the bricks in their weather-proof covering in a sheltered area.

◆ When bales of bricks are opened and are not completely used up, cover the loose bricks with a damp-proof material.

◆ If the bricks become damp, allow them to dry out before use.

◆ Protect uncompleted brickwork at night and during rain. One way to do this is to turn back the scaffolding boards against the brickwork to protect the brickface and cover the top of the wall with a damp-proof material.

◆ Avoid saturating the brickwork for cleaning purposes.

◆ Ensure that all points of the building which are likely to be exposed to continuous saturation are protected by adequate damp-proof courses.

If these precautions are followed, and the correct type of brick for the particular climatic conditions is chosen, the possibility of efflorescence is minimised.

REMOVAL OF EFFLORESCENCE

Efflorescence should be allowed to weather away naturally, but it may be removed by brushing with a stiff bristle brush. The deposit should be collected and removed so that it does not enter the masonry at lower levels. Any deposit remaining may be removed or reduced by treatment with clean cold water. Since the deposit is water-soluble, washing down may result in the solution being partially re-absorbed. This may be minimised by using a clean damp sponge, which should be rinsed frequently in clean water. Recurrent efflorescence on older established brickwork may almost always be taken as an indication that considerable quantities of water are entering the masonry as a result of failure of weathering and other protective measures, faulty spouts and gutters and the like.

Chemical methods should never be used for the removal of efflorescence.

Note: I.S. 51 "Clay Building Bricks" 1983 outlines a test for determining amounts of efflorescence, and clarification can be sought from the manufacturer of the possibility of the amount of efflorescence as determined from that test.

This standard also defines the permissible amount of soluble salt content allowed and the manufacturer should be able to confirm that bricks supplied are in accordance with the limits set down.

SULPHATE ATTACK

Incidents of sulphate attack are rare in Ireland. It is caused by a reaction of chemicals in the bricks with chemicals in the mortar solution. The tricalcium aluminate in ordinary Portland cements reacts with the brick sulphate and causes expansion. The source of sulphate is usually the soluble salts present in varying extents in clay bricks, but may also be derived from ground water and from flue gases.

As stated earlier incidents of sulphate attack are rare and are only serious when the design of the building allows continuous saturation of the brickwork. The reaction is firstly an overall expansion of the brickwork, followed, in more extreme cases, by a progressive disintegration of the mortar joints.

The three causes of sulphate attack are tricalcium aluminate in Portland cement, soluble sulphate and water. The possibility of attack should be considered at the design stage of the building and combination of mortars and brick types susceptible to sulphate erosion should be avoided.

Tricalcium aluminate is present in all ordinary Portland cements in amounts from eight to thirteen percent. There is no easy way to estimate the susceptibility of different brand-name cements, so it must be assumed that all ordinary Portland cements are capable of being eroded.

There are a number of ways to safeguard against attack:

◆ Specify richer mortar mixes, for example, $1:\frac{1}{4}:3$ or $1:\frac{1}{2}:4\frac{1}{2}$ or (better still) 1:5-6 with plasticiser in place of lime.

◆ Use sulphate-resisting Portland cement.

◆ Use bricks with low water absorption percentage.

◆ Use bricks with low soluble salt content.

Most types of clay bricks contain soluble sulphates; the amounts vary between different types of bricks. Special quality bricks have limitations on their sulphate content but ordinary bricks have no such limitations. It therefore must be assumed that the use of ordinary bricks will contribute to sulphate attack.

Repeated wetting and drying out of brickwork over a period of years is closely related to sulphate attack. The susceptibility also depends on the exposure to wet weather. Exposed brickwork such as parapets and free-standing walls are most likely to be affected and consideration for this should be made in the building design.

FROST ATTACK ON BRICKWORK

In Ireland frost damage is rare in normal external walls. Trouble can occur, however, in walls exposed to both frost and rain. Frost attack occurs only when brickwork is saturated as well as frozen. So brickwork is more vulnerable in parapets and freestanding and retaining walls where exposure conditions cause continual saturation.

Precautions against frost attack should be taken when choosing the quality of the bricks and mortars.

In extremely exposed conditions, it may be necessary to use special quality bricks which have high compressive strengths and a low water absorption rate. Engineering bricks fulfil these requirements and have known water absorption rates. There is no standard for water absorption rates for ordinary bricks; these can be found only by approaching each manufacturer and asking for results of tests carried out on their bricks.

Choosing a special quality brick could impose limitations on the range of brick colours and textures and this effect should be considered in the choice of materials for all other parts of the building. Despite intensive research over many years, there is no presently accepted test for frost resistance. The only requirement that the brick manufacturer must satisfy is evidence that bricks of the type required have given satisfactory results for at least three years in the locality concerned in conditions at least as severe as those proposed. When a suitable brick has been found, one precaution would be to examine all the buildings using this brick in the same weather conditions. Frost attack is likely to affect a building within the first three years, so buildings older than this using the same type of brick would be a good reference.

In particular if bricks which have not been previously used are proposed or if the area has a high incidence of freezing and driving rain, it would be advisable to obtain confirmation from the manufacturer that the bricks are suitable for use in that location.

Frost attack on brickwork.

GLAZING

INTRODUCTION

Technical Guidance Document D "Materials and Workmanship" of the Building Regulations require that all works to which Building Regulations apply shall be carried out with proper materials and in a workmanlike manner.

With respect to glazing, various codes of practice exist which are referred to in the following text.

Thickness of glass

If the location of the glass is such that there is a risk of accidental breakage which may result in injury, specific recommendations exist which are referred to in this section. In other locations (non-risk areas) the quality and thickness of normal window glass should be specified to suit the design wind load for the location of the dwelling.

A number of factors affect the thickness of glass to be used. These include wind speed, site topography, roughness of ground, height of eaves above ground and size and shape of pane of glass. As some of the factors change in locations across the country it is not possible to give a single specific requirement for thickness of glass for a specific size of pane.

British Standard BS 6262 Code of Practice CP 3, Chapter V Part 2:1972 "Basic Data for the Design of Building: Loadings" indicate how the wind loading may be determined. An abbreviated method of determining glass thickness for the majority of low rise housing is contained in BS 6262. Relevant parts of this document are reproduced later in this section.

GLASS THICKNESS CALCULATION METHOD, FOR NON-RISK AREAS

1   Obtain wind speed for the location of the dwelling, from wind speed map on page 272.

2   From tables 1 and 2 page 273, obtain design wind pressure.

3   From the transparent annealed glass graph on page 274, determine thickness required (knowing wind loading and area of glass). In this graph the shaded area for each thickness relates to aspect ratio, which is the ratio of length to breadth of pane of glass. The bottom line of the shaded area refers to aspect ratio 1 : 1 (i.e. square) and top line 3 : 1.

In calculating the thickness from the graph the point where the vertical line for the required wind loading intersects the horizontal line for the required area is determined. If the point of intersection is in the unshaded area, the next greater thickness of glass is required. Should the intersection point be within a shaded band, determine the aspect ratio to which that point corresponds by interpolation between aspects ratios of 1 :1 lower limit and 3 : 1 upper limit of shaded band. If the value of the calculated aspect ratio is nearer the thick black line (lower) than the interpolated aspect ratio or equal to the interpolated aspect ratio, the next greatest thickness of glass should be used. Otherwise the glass thickness corresponding to the shaded band may be used. Page 270 gives a visual interpretation of this procedure.

INTERPOLATED ASPECT RATIO

The extract from the transparent annealed glass graph reproduced below illustrates how aspect ratio is interpolated.

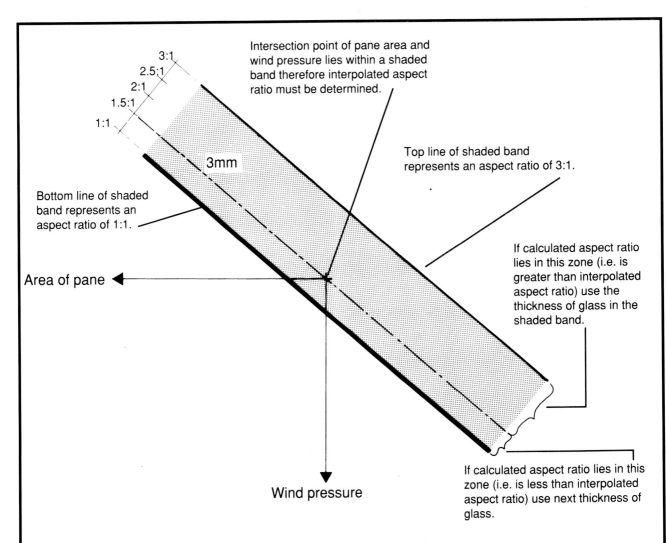

3:1
2.5:1
2:1
1.5:1
1:1

Intersection point of pane area and wind pressure lies within a shaded band therefore interpolated aspect ratio must be determined.

Top line of shaded band represents an aspect ratio of 3:1.

Bottom line of shaded band represents an aspect ratio of 1:1.

3mm

Area of pane

If calculated aspect ratio lies in this zone (i.e. is greater than interpolated aspect ratio) use the thickness of glass in the shaded band.

Wind pressure

If calculated aspect ratio lies in this zone (i.e. is less than interpolated aspect ratio) use next thickness of glass.

Example 1
If the calculated aspect ratio of a window pane is 1:1 and from the sketch above the interpolated aspect ratio was determined as 1.5:1, then the next thickness of glass indicated should be used, i.e., 4mm.

Example 2
If the calculated aspect ratio of a window pane is 2:1 and from the sketch above the interpolated aspect ratio was determined as 1.5:1, then the thickness of glass indicated in the shaded band may be used, i.e., 3mm.

**GLASS THICKNESS CALCULATION
METHOD, FOR NON-RISK AREAS**
continued

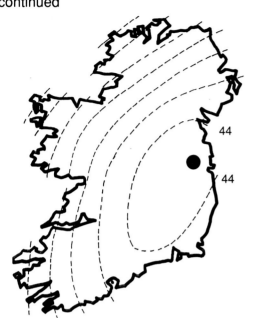

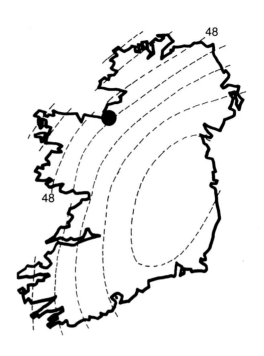

Example 1.

Two story house
Location: Kildare
Window pane size: 1m x 1m
Window pane area: 1m²
Actual aspect
ratio of window pane: 1:1

◆   From wind speed map, Kildare = 44m/sec.

◆   From Table 1 – Ground roughness
    categories, assume condition 1.

◆   From Table 2 determine design wind
    pressure, assume eaves height of 5m =
    1300N/m² wind pressure.

◆   From the transparent annealed glass graph,
    draw a vertical line up from 1300 N/m² and
    horizontally across from 1m². As these lines
    intersect in a shaded band, representing
    3mm we must establish the interpolated
    aspect ratio of the window pane which is
    approximately 2:1. As the calculated aspect
    ratio of the pane (1:1) is less than this then
    the next thickness of glass must be used,
    i.e., 4mm.

Example 2.

Two story house
Location: Sligo
Window pane size: 1m x 1m
Window pane area: 1m²
Actual aspect
ratio of window pane: 1:1

◆   From wind speed map, Sligo = 48m/sec.

◆   From Table 1 – Ground roughness
    categories, assume condition 1.

◆   From Table 2 determine design wind
    pressure, assume eaves height of 5m =
    1550N/m² wind pressure.

◆   From the transparent annealed glass graph,
    draw a vertical line up from 1550N/m² and
    horizontally across from 1m². As these lines
    intersect in an unshaded band the next
    thickness of glass should be used, i.e.,
    4mm.

**GLASS THICKNESS CALCULATION
METHOD, FOR NON-RISK AREAS**
continued

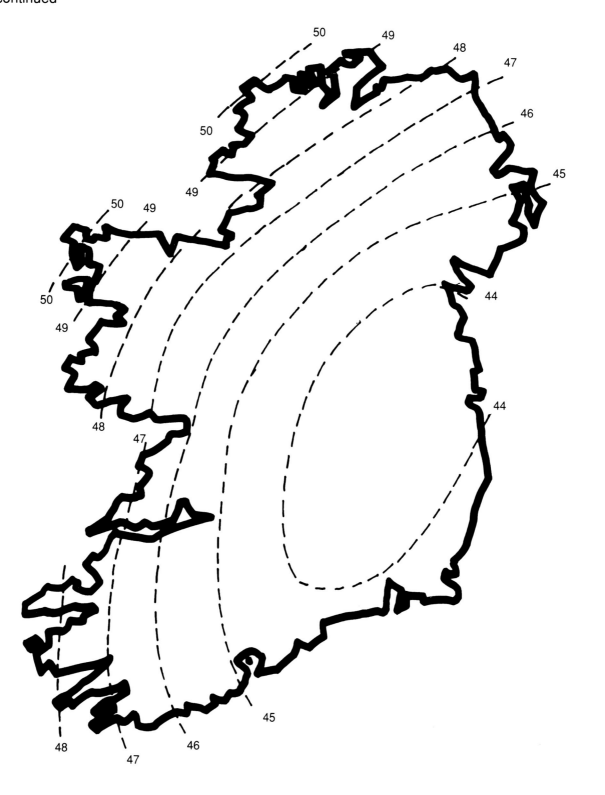

Estimated maximum gust speed (M/S) with
return period 50 years. Valid for height of 10m
above open level country.

Wind speed map

**GLASS THICKNESS CALCULATION
METHOD, FOR NON-RISK AREAS**
continued

TABLE 1 GROUND ROUGHNESS CATEGORIES	
Ground description	**Category**
Long fetches of open, level or nearly level country and all coastal situations	1
Open country with scattered windbreaks	2
Country with many windbreaks; small towns: outskirts of large cities	3
Surfaces with large and frequent obstructions, e.g. city centres	4

TABLE 2		DESIGN WIND PRESSURES (N/m²)			
Basic wind speed (m/s)	Height to eaves	Design wind pressure for a ground roughness category. (See table 1)			
		Cat. 1	Cat. 2	Cat.3	Cat.4
50	3	1500	1150	900	700
	5	1700	1350	1050	800
	10	2150	1850	1300	1000
48	3	1400	1050	850	650
	5	1550	1250	1000	750
	10	2000	1750	1200	900
46	3	1250	950	750	600
	5	1400	1150	900	650
	10	1850	1600	1100	850
44	3	1150	900	700	600
	5	1300	1050	850	600
	10	1200	1450	1050	750

**GLASS THICKNESS CALCULATION
METHOD, FOR NON-RISK AREAS**
continued

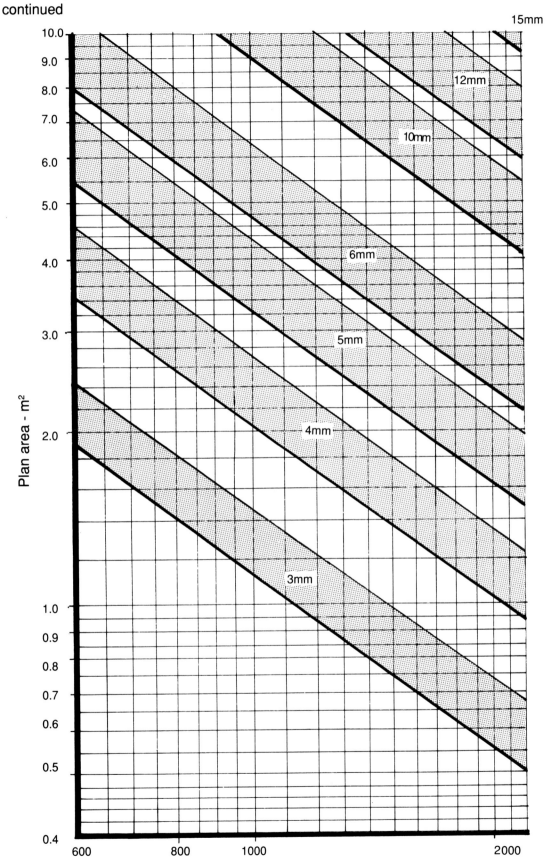

Design wind pressure – N/m² (from table 2)

Transparent annealed glass

SAFETY GLAZING IN DOORS AND SIDE PANELS

In doors, side panels and any area where there is a high risk of accidental breakage, the glazing should be designed and selected to comply with the safety recommendations for risk areas, as specified in British Standard BS 6262 "Code of Practice for Glazing for buildings".

Where there is a particular risk, such as at door side panels, "low level" glazing and where fully glazed panels can be mistaken for doors, toughened or laminated glass or other materials may be needed.

The code recommends that fully glazed doors, e.g. patio doors and ordinary doors where the glazing takes up most of the door, should use a safety glazing material to BS 6206 and of the appropriate class.

Doors with more than one pane or where a single pane does not take up most of the door can be glazed in 6mm or thicker annealed (normal) glass.

Where a door side panel could be mistaken for a door these recommendations also apply.

The sketches of various door types show where safety glazing should be used and where 6mm or thicker annealed glass may be used.

Note: Bent glass and glass bullions are excluded from the code.

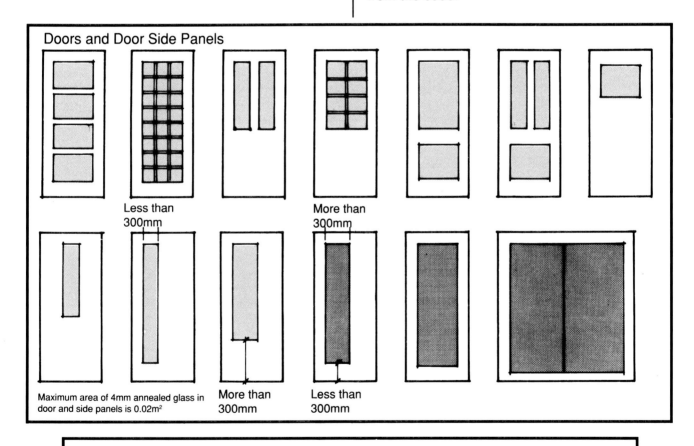

Doors and Door Side Panels

Less than 300mm

More than 300mm

Maximum area of 4mm annealed glass in door and side panels is 0.02m²

More than 300mm

Less than 300mm

Not less than 6mm
Annealed Glass
Note: Areas of glass less than 0.02m² can be 4mm thick glass

Safety Glazing Material
Doors and side panels over 900mm wide not less than Class B
Doors and side panels over 900mm wide not less than Class C low level glazing mostly Class C

This table is reproduced by kind permission of NHBC England. The source of this material is NHBC and BSI.

LOW LEVEL GLAZING

The Code recommends that in places where many people, especially children, are moving about e.g. corridors, landings, bottom of stairs, glazing that comes lower than 800mm (31") from the floor should be safety glazing. Where glazing is protected by a barrier rail, annealed glass within the limits set by the table (reproduced here) may be used.

In more usual situations where few people are likely to be moving about, annealed glass within the limits set by the table may be used. The sketches show typical examples of these recommendations.

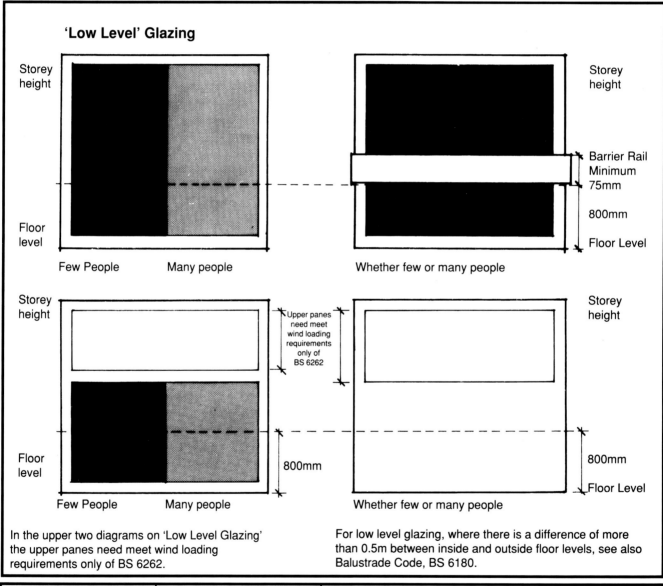

'Low Level' Glazing

In the upper two diagrams on 'Low Level Glazing' the upper panes need meet wind loading requirements only of BS 6262.

For low level glazing, where there is a difference of more than 0.5m between inside and outside floor levels, see also Balustrade Code, BS 6180.

Single Glazing		Hermetically Sealed, Factory Made Insulating Glass Units	
Nominal Thickness mm	Maximum Area m²	Nominal Thickness mm	Maximum Area m²
4	0.2	4+4	0.6
5	0.8	5+5	1.2
6	1.8	6+6	2.5
10	3.3	10+10	5.0
12	5.0		

Safety Glazing Material
Doors and side panels over 900mm wide not less than Class B
Doors and side panels over 900mm wide not less than Class C
low level glazing mostly Class C

See Table

DOUBLE – GLAZED UNITS

BS 6262 can be used to determine the thickness of glass for double glazed units and the manufacturer of the units should make reference to same knowing the location of the dwelling. British Standard BS 5713 "Double Glazing" makes reference to the design and installation of double glazing.

Summary
In all cases the designer should ensure that the builder of the house and the glazing contractor specify and install the correct thickness of glass as required in the relevant codes.

Acknowledgement
Extracts from BS 6262: 1982 are reproduced with the permission of BSI. Complete copies can be obtained by post from National Standards Authority of Ireland, Glasnevin, Dublin 9, or BSI Sales, Linford Wood, Milton Keynes, MK14 6LE, England.

SITE SAFETY

SAFETY ON SITE

Over 70,000 people are employed in construction – either on building sites or repair work. It is very easy to be seriously injured or killed at work, yet there is always a safe way to do any job – in building it is also the quickest and cheapest.

The simple message of this section is that it is easy to take the right precautions.

Members are advised that new Health and Safety regulations to supplement the EC temporary or mobile construction sites directive 92/57/EEC will be enforced during 1994. This deals with minimum Health and Safety standards on sites and places additional duties on clients, designers and contractors. Members are referred to the Council Directive 92/57/EEC available from Government Publication Sales Office, Sun Alliance House, Molesworth Street, Dublin 2.

Securing a ladder

Lethal ladders

Incorrect use of ladders kills a lot of people. Make sure the ladder is:

◆ Right for the job. Would scaffolding be better?

◆ In good shape.

◆ Secured near the top.

◆ On a firm base and footing.

◆ Rising at least 1 metre beyond the landing place or that there is a proper hand hold.

◆ Always have a firm grip on the ladder and keep a good balance.

NEVER

◆ Use a makeshift ladder.

◆ Lean sideways from a ladder.

SCAFFOLDING

Working at heights is a serious job. When using scaffolding be very careful; apply the following common sense guidance illustrated here and on the next page.

Guard rails and toe boards

Scaffold platforms from which a person may fall more than 2m must have guard rails and toe boards. Brick guards or other suitable vertical protection should be provided where materials may fall from the scaffold (though scaffold fans, netting or sheeting may be more effective in certain circumstances).

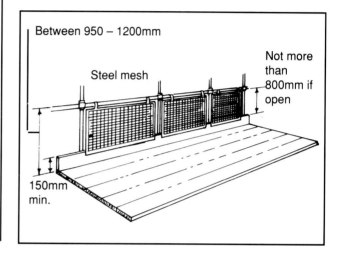

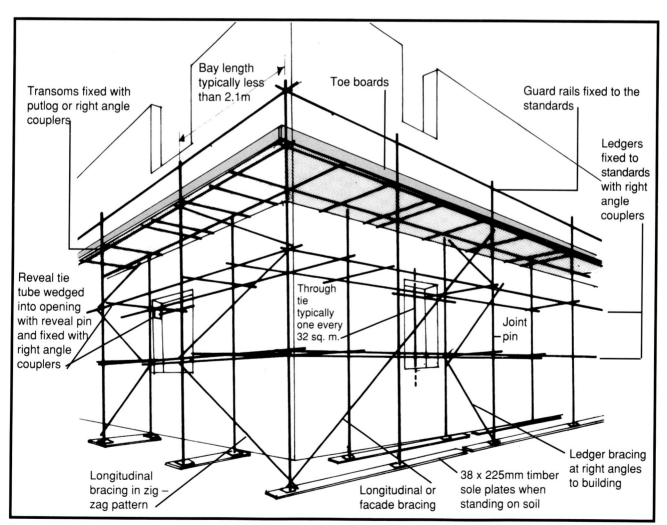

Typical independent tied scaffold (general duty – unsheeted)

GUIDELINES FOR SCAFFOLDING

◆ Is there proper access to the scaffold platform?

◆ are all uprights provided with base plates (and, where necessary, timber sole plates) or prevented in some other way from slipping or sinking?

◆ have any uprights, ledgers, braces or struts been removed? If so, have replacements been provided?

◆ is the scaffold secured to the building in enough places to ensure stability?

◆ if any ties have been removed since the scaffold was erected have substitute ties been provided to maintain stability?

◆ are the working platforms fully boarded?

◆ are there effective barriers or warning notices to stop people using an incomplete scaffold, e.g. one that isn't fully boarded?

◆ are there adequate guard rails and toe boards at every side from which a person can fall and in particular where one can fall more than 2.0m (6ft. 6in.)?

◆ does the working platform extend at least 600mm beyond the end of the working face?

◆ where the scaffold has been designed and constructed for loading with materials, are these evenly distributed?

◆ are all wheeled scaffolds used only on firm and even surfaces?

◆ are all suspended scaffolds

—closely boarded or planked,

—at least 600mm wide if used as footing only,

—at least 800mm wide if used for materials?

◆ are trestle scaffolds used only on level ground and for light work of a short duration?

◆ does a competent person inspect the scaffold before it is brought into use and on a regular basis thereafter, i.e. at least once a week and always after bad weather?

◆ are the results of inspections recorded (including defects that were put right during the inspections) and the records signed by the person who carried the inspections?

Boards and Planks on Working Platforms

◆ are boards free from obvious defects such as knots, and are they arranged to avoid tipping or tripping?

◆ are all boards and planks on working platforms

— at least 200mm wide if less than 50mm thick,

—at least 150mm wide if more than 50mm thick?

◆ do these planks extend beyond their end support a distance of at least 50mm and not more than four times their thickness?

◆ The distance between transoms should not be more than

—1 metre where planks of 32mm thickness are used,

—1.5 metres where planks of 38mm thickness are used,

—2.4 metres where planks of 50mm thickness are used.

THE ROOF: A RISKY PLACE TO BE

So risky, in fact, that one in five deaths on sites arise from roof work. Many of these accidents happen in the course of routine maintainence. Almost all of them could be prevented. Very simply.

◆ Always inspect a roof before walking on it.

◆ Use proper roofing ladders/crawling boards on sloping or fragile roofs.

◆ If there are others working underneath the roof, make sure debris cannot fall on them.

Excavations

◆ All excavations deeper than 1.25m must be shored or sloped back to a safe angle.

◆ Before digging, make sure that the location of underground pipes and services is first established.

General

◆ One of the most dangerous places in a building under construction is around the stairwell and lift shafts in apartments. Make sure that the area is well protected, and that temporary guard rails and balustrades are in place to prevent people falling down.

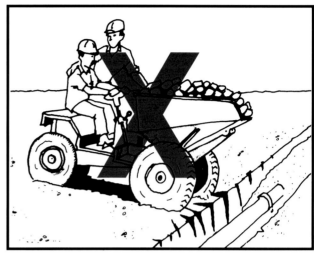

◆ Don't carry passengers in a vehicle that is not designed for passengers.

◆ Always ensure nails are removed or hammered down.

◆ Safety helmets must be worn on all construction sites.

◆ Hard–toed footwear should be worn.

◆ Goggles, ear defenders and gloves may be necessary.

ELECTRICITY, A KILLER WHEN MISUSED ON SITE

Electrical accidents, many of which are fatal, are often caused by contact with:

◆ underground or overhead power lines

◆ unsuitable or badly maintained equipment

◆ bad connections to the supply

Here's how to handle electricity on site:

◆ treat electricity with respect

◆ check constantly that cables are not damaged or worn

◆ keep trailing cables off ground and away from water

◆ never overload or use makeshift plugs and fuses

◆ for mains voltage, screened cables must be used and circuits must be protected by proper circuit breakers.

PORTABLE POWER TOOLS... PORTABLE POWER DANGER

◆ always use the right tool for the job, and don't make do with a defective tool.

◆ check all tools before use (ensure they are properly earthed)

◆ don't adjust power tools unless the supply is disconnected

◆ always be careful of angle grinders and power saws, and check suitable guards are fitted and used

◆ they must operate at a reduced voltage (110V)

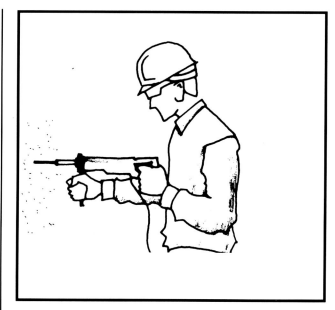

Cartridge operated tools

◆ always follow maker's instructions

◆ keep in secure place when not in use

◆ you should be trained to use all tools

◆ if working on hard material wear goggles, don't be blind to the risk.

LIFTING BY HAND CAN DAMAGE YOUR BACK

Lifting weights that are too heavy for you, or just lifting weights the wrong way, will do your back permanent damage. You may feel the damage straight away or, more likely the back pain will show up over time. It's very easy to avoid this back damage. It just takes commonsense.

◆ get a good grip, keep the load close to your body

◆ keep the back straight

◆ bend your knees, lift with your leg muscles not your back

◆ if it's too heavy, get help

◆ start lifting sensibly, if you don't want to end up permanently disabled with a bad back before you're 30.

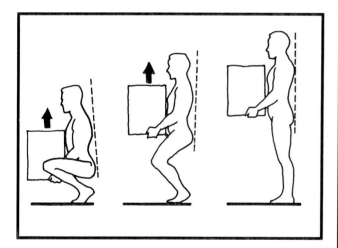

Additional Information.

Additional information on all aspects of site safety is available from:

Health and Safety Authority (HSA) and National Irish Safety Organisation (NISO).
10 Hogan Place,
Grand Canal St.,
Dublin 2.

The detailed legal requirements in respect of site safety are set out in the:

◆ Construction (Safety Health and Welfare Regulations (SI 282 of 1975),

◆ Safety Health and Welfare at Work Act, no. 7 of 1989.

◆ Safety Health and Welfare at Work (General Application) Regulations 1993 (SI 44 of 1993).

Each of the above is available from:
Government publications,
Sales Office,
Sun Alliance House,
Molesworth St.,
Dublin 2.

NOTE: For any aspects of site safety or safety in the work place not covered by the above publications additional information can be obtained from Government Publications Sales Office or the Health and Safety Authority.

SOUND

INTRODUCTION

Sound is a form of energy which can be transmitted over a distance from its source through a medium, such as air or a solid element of construction e.g. a wall or a floor. The types of sound to be considered are airborne and impact sounds. In each case the sound may be transmitted directly or indirectly, see sketch below.

The principal methods of isolating the receiver from the source are:

◆ Eliminating pathways along which sound can travel, and

◆ Using barriers formed of materials of sufficiently high mass which will not easily vibrate.

In practice sound insulation is usually achieved by using a combination of both methods.

Note:

◆ There are no sound requirements for detached houses.

◆ Sound requirements for semi–detached and terraced houses relate mainly to the party wall.

◆ There are no sound requirements for floors in single family dwelling houses.

◆ The information in this Appendix is for single family dwelling houses, and reference should be made to Technical Guidance Document E of the Building Regulations for sound insulation requirements for flats and apartments.

Transmission of sound

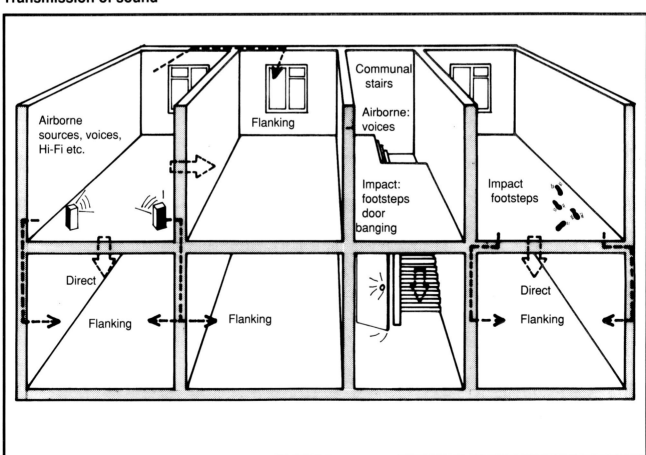

TRANSMISSION OF SOUND

As can be seen from the sketch on the previous page there are two sources of sound energy:

◆ Airborne source: voices, hi–fi, tv's etc.,

◆ Impact source: footsteps, doors banging, etc.,

Both these types of sound can be transmitted directly or indirectly.

Direct transmission of sound

Direct transmission means the transmission of sound directly through a wall or floor from one of its sides to the other.

Walls should reduce the level of airborne sound. A solid masonry wall depends on its mass; being heavy, it is not easily set into vibration. Walls with two or three leaves depend partly on their mass and partly on structural isolation between the leaves.

With masonry walls the mass is the main factor but stiffness and damping are also important. Cavity masonry walls need at least as much mass as solid walls because their lower degree of stiffness offsets the benefit of isolation.

Air–paths must be avoided, porous materials and gaps at joints in the structure must be sealed. Joints in blockwork in party walls should be filled with mortar. It is preferable that party walls should also have a wet plaster finish to eliminate air–paths from sound transmission through the wall. Resonances must also be avoided; these may occur if some part of the structure (such as a dry lining) vibrates strongly at a particular sound frequency (pitch) and transmits more energy at this pitch.

Flanking transmission of sound

Flanking transmission means the indirect transmission of sound from one side of a wall or floor to the other side.

Because a solid element may vibrate when exposed to sound waves in the air on both sides. Flanking transmission happens when there is a path along which sound can travel between elements on opposite sides of a wall or floor. This path may be through a continuous solid structure or through an air space (such as the cavity of an external wall). Usually paths through structure are more important with solid masonry elements, where paths through an air space are more important with thin panels (such as studwork and ceilings) in which structural waves do not travel as freely.

WALLS AND FLOORS

The location of walls and floors which are required by regulations to have good sound insulation.

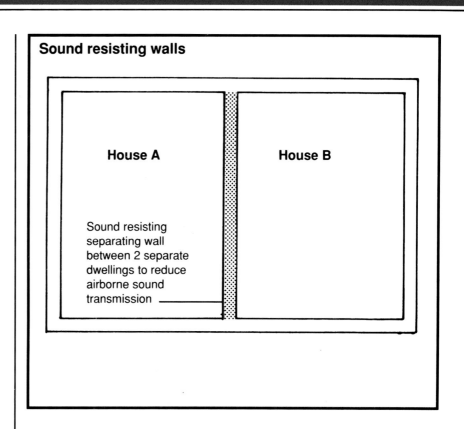

Sound resisting walls

House A

House B

Sound resisting separating wall between 2 separate dwellings to reduce airborne sound transmission

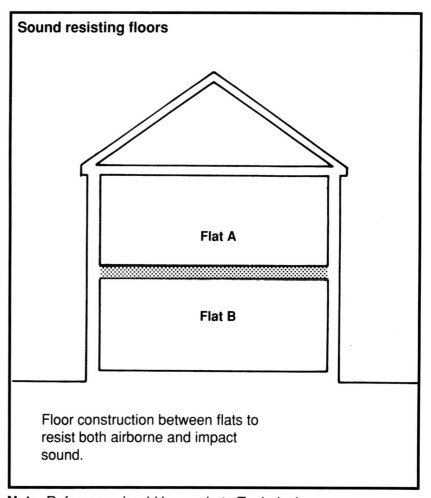

Sound resisting floors

Flat A

Flat B

Floor construction between flats to resist both airborne and impact sound.

Note: Reference should be made to Technical Guidance Document E of the Building Regulations for the detailed sound requirements between flats and apartments.

TYPES OF WALL

One of the methods of reducing sound transmission is to use materials in the wall of sufficiently high mass that they will not vibrate easily. The two more widely used wall constructions are:

1) Solid masonry party wall

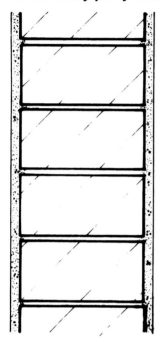

Concrete blockwork or concrete brickwork plastered on both faces

The average mass of the wall (including the plaster) should be at least 415kg/m². The thickness of the plaster should be at least 12.5mm on each face. Use blocks which extend to the full thickness of the wall.

Example:
215mm concrete block, 112.5mm coursing, lightweight plaster; block density 1860kg/m³ gives the required mass.

Point to watch
Fill the joints between the bricks or blocks with mortar, and seal the joints between the wall and the other parts of the construction (to achieve the mass and to avoid air paths).

Limit the pathways between the walls and opposite sides of the sound resisting wall (to reduce flanking transmission).

2) Cavity masonry party wall

Two leaves of concrete blockwork or concrete brickwork plastered on both faces

The width of the cavity should be at least 50mm. The average mass of the wall (including the plaster) should be at least 415kg/m².
The thickness of the plaster should be at least 12.5mm on each face.

Example:
102mm leaves, 225mm coursing, lightweight plaster; block density of 1961 kg/m³ gives the required mass.

Points to watch
Fill the joints between the bricks or blocks with mortar, and seal the joints between the wall and the other parts of the construction (to achieve the mass and to avoid air paths).

Maintain the separation of the leaves and space them at least 50mm apart. Connect the leaves with butterfly pattern wall ties.

If a cavity in an external wall is completely filled with an insulating material other than loose fibre, care should be taken that the insulating material does not enter the cavity in the separating wall.

key junctions in the construction.

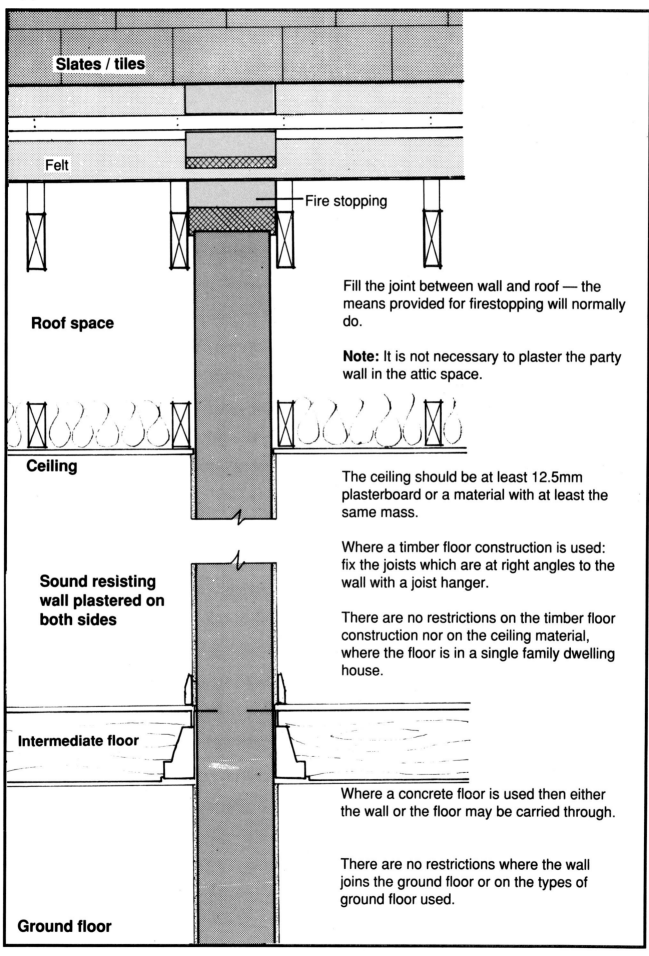

Slates / tiles

Felt

Fire stopping

Roof space

Fill the joint between wall and roof — the means provided for firestopping will normally do.

Note: It is not necessary to plaster the party wall in the attic space.

Ceiling

The ceiling should be at least 12.5mm plasterboard or a material with at least the same mass.

Sound resisting wall plastered on both sides

Where a timber floor construction is used: fix the joists which are at right angles to the wall with a joist hanger.

There are no restrictions on the timber floor construction nor on the ceiling material, where the floor is in a single family dwelling house.

Intermediate floor

Where a concrete floor is used then either the wall or the floor may be carried through.

There are no restrictions where the wall joins the ground floor or on the types of ground floor used.

Ground floor

KEY JUNCTIONS IN THE CONSTRUCTION
continued

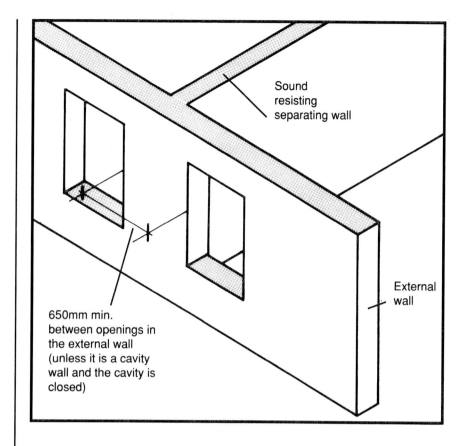

Sound resisting separating wall

External wall

650mm min. between openings in the external wall (unless it is a cavity wall and the cavity is closed)

There should be at least 650mm between openings in the external wall (unless it is a cavity and the cavity is closed)

If the external wall is of cavity construction, there are no restrictions on a masonry outside leaf.

Masonry external wall (either a solid wall or the inner leaf of a cavity wall) should be either bonded to the sound resisting wall or butted to it and secured with wall ties (or similar) spaced no more than 300mm apart vertically.

DRAINAGE

INTRODUCTION

The following section deals with drainage above and below ground and all the information contained is in accordance with Technical Guidance Document H of the Building Regulations.

DEFINITIONS.

Surface water: Is the run-off of rainwater from roofs and any paved surface around the building.

Soil water: Is water containing excreted material i.e. from W.C. pans.

Waste water: Is used water from waste appliances i.e. sink baths, showers, WHB's. Used water from washing machines and dishwashers is waste water.

Foul water: Is any water contaminated by soil water or waste water.

TRAPS

All points of discharge into the drainage system should be fitted with a water seal (trap) to prevent foul air from the system entering the building.

Table 1 gives minimum trap sizes and seal depths for the most commonly used appliances.

Ventilation — To prevent the water seal from being broken by the pressures which can develop in the system the branch discharge pipes should be designed as described on page 297.

Access for clearing blockages — If a trap forms part of an appliance as in a W.C. pan the appliance should be removable.
All other traps should be fitted to the appliance and should be removable or be fitted with a cleaning eye.

Table 1 Minimum trap sizes and seal depths.

Appliance	Diameter of trap (mm)	Depth of seal (mm)
Washbasin, Bidet	32	75
Sink*, Bath*, Shower*, Food waste disposal unit, Urinal bowl	40	75
W.C. pan	75	50

* Where these appliances are installed on a ground floor and discharge to a gully, the depth of seal may be reduced to not less than 40mm.

Typical trap

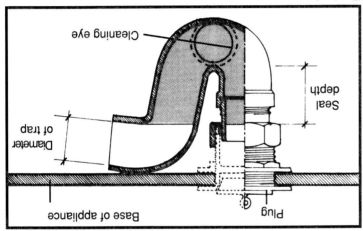

Cleaning eye

Seal depth

Diameter of trap

Base of appliance

Plug

Typical domestic drainage: definitions

Foul water discharge stack

House drain

Surface water rain water pipe discharging to gully

Waste water branch discharge pipe to gully

Waste water branch discharge pipe

Soil water branch discharge pipe from W.C.

waste water branch discharge pipe from W.C.

Vent stack

Surface water gutter

BRANCH DISCHARGE PIPES

Branch discharge pipes should satisfy the requirements on the following pages as applicable.

1 Branch pipes should discharge into another branch discharge pipe or a discharge stack unless the appliances are on the ground floor.

2 If the appliances are on the ground floor the branch pipe(s) may discharge to a discharge stack or directly to a drain, or (if the pipe carries only waste water) to a gully.

3 Junctions on branch pipes should be made with a sweep of 25mm minimum radius or at 45°. Connection of branch pipes of 75mm diameter or more to the stack should be made with a sweep of 50mm minimum radius or at 45°.

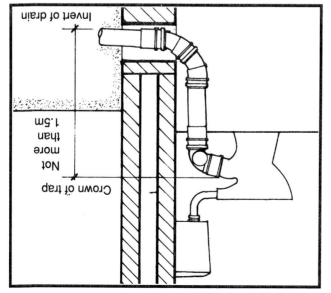

4 A branch pipe from a ground floor water closet should only discharge directly to a drain if the drop is less than 1.5m as illustrated.

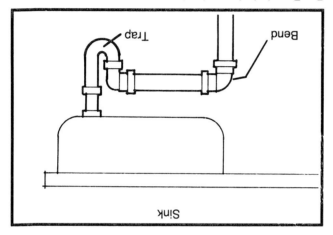

5 Bends in branch pipes should be avoided if possible but where they are essential they should have a radius as large as possible. Bends on pipes of 65mm diameter or less should have a centreline radius of at least 75mm.

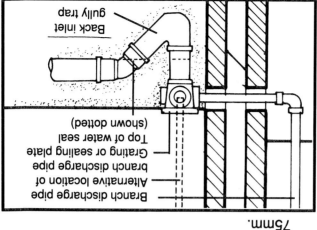

6 A branch pipe discharging to a gully should terminate between the grating or sealing plates and the top of the water seal.

BRANCH DISCHARGE PIPES, continued.

7   Ventilated branch pipes — where separate ventilation is not provided, the length and slope of the branch discharge pipes should not exceed those shown in figures 1 and 2, in order to prevent the water seals in traps from being lost by pressure which can develop in the system.

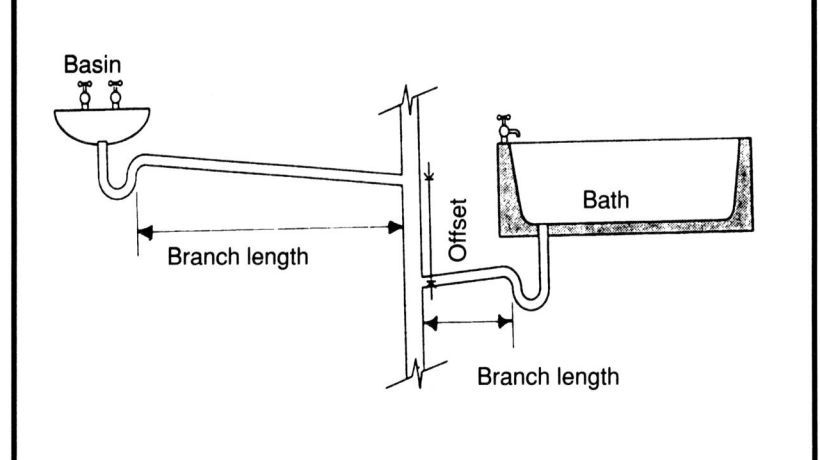

Washbasin waste pipes
◆   Trap 32 min. diameter and 75 min. seal.

◆   WHB branch discharging to stack:
    1.7m max. branch length for 32 dia. pipe, gradient as per design curve below. 3.0m max. for 40 dia. pipe.

Offsets:
◆   Offset all small sized connections (up to 65mm) opposite each other by:
    110mm min. on a 100mm dia. stack.
    200mm min. on a 150mm dia. stack.
    See figure 2, page 299, for offsets at larger connections.

Sink, Bath or shower.
◆   Trap 40 min. dia .

◆   40 min. seal if on ground floor and discharging to gulley, otherwise 75mm min. seal.

◆   Sink or bath branch discharging to stack:
    3.0 max. branch length for 40 dia. pipe.
    Slope between 18 and 90mm/m.

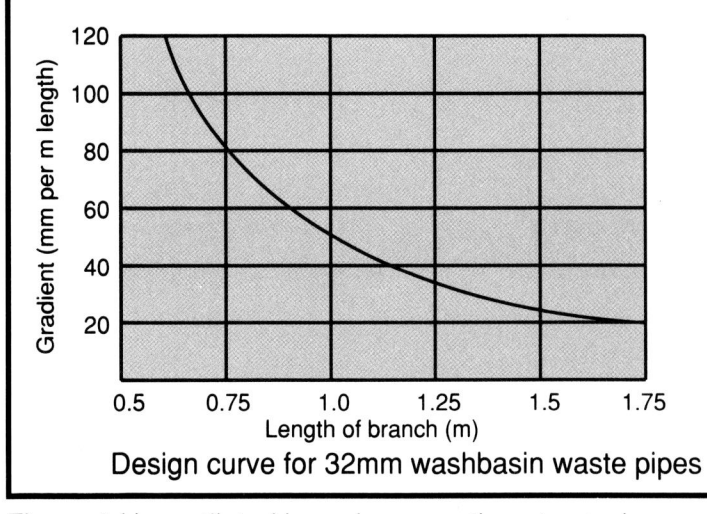

Design curve for 32mm washbasin waste pipes

Figure 1 Unventilated branch connections to stack

BRANCH DISCHARGE PIPES
continued.

VENTILATED BRANCH PIPES

If the guidance given in figures 1 and 2 cannot be achieved the branch discharge pipe should be ventilated by a branch ventilating pipe. This is most commonly achieved by connecting the branch ventilating pipe to the discharge stack.

Branch ventilation pipes should be connected to the branch discharge pipe within 300mm of the trap and should not connect to the stack below the "spillover" level of the highest appliance served (see sketch opposite).

The ventilation pipe should have a continuous incline from the discharge pipe to the point of connection to the stack.

Branch ventilation pipes serving one appliance should be at least 25mm min. diameter or where the branch is longer than 15m or has more than 5 bends, should be at least 32mm.

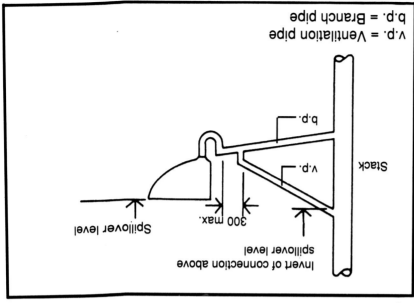

Figure 3 Ventilated branch pipe.

v.p. = Ventilation pipe
b.p. = Branch pipe

Diagram labels: Stack — Spillover level — Invert of connection above spillover level — 300 max. — b.p. — v.p.

W.C.

- 75 min. diameter trap, 50 min. seal depth
- Max W.C. branch length 6.0m long for single W.C.
- Sweep W.C. branch connection to stack min. 50mm radius.
- Where a waste enters the stack opposite and within 200 below W.C. branch centreline, make an angled connection (as shown dotted above) or a parallel junction between waste and stack as illustrated here.
- Branch pipe slope 9mm/m min.

Note:
Where the branch pipe diameter to a sink or washbasin is greater than the appliance outlet diameter the following applies.

1 Trap diameter to be the same as the appliance outlet.

2 Extend tail of trap by 50mm before connecting to the larger diameter branch pipe.

Note: Rodding points should be provided to give access to any lengths of discharge pipes which cannot be reached by removing traps.

Figure 2 Branch connections to stack.

Diagram labels: Appliance — Pipe gradient as per Fig 1 — 200mm — W.C. — Lowest connection to stack to be min. 450 above invert at stack base.

DISCHARGE STACKS

Discharge stacks should satisfy the following requirements as applicable.

1. All stacks should discharge to a drain. The bend at the foot of the stack should have as large a radius as possible and should be at least 200mm at the centre line. See sketch opposite.

2. Offsets in the 'wet' portion of the discharge stack should be avoided. If they are unavoidable then, in a building of not more than 3 storeys, there should be no branch connection within 750mm of the offset. Bends or offsets are permitted in the 'dry' portion of the stack.

3. Sizes generally in housing are 100mm min. diameter. The only exception to this is where the W.C. connection is from a single syphonic closet. The dry portion can also be reduced to 75mm diameter in one or two storey houses.

4. Ventilation of discharge stacks is necessary to prevent water seals in traps from being lost by pressures which can develop in the system. Discharge stacks must be ventilated.

5. Rodding points should be provided to give access for clearing blockages to any lengths of pipe which cannot be reached from any other part of the system. All pipes should be reasonably accessible for repair.

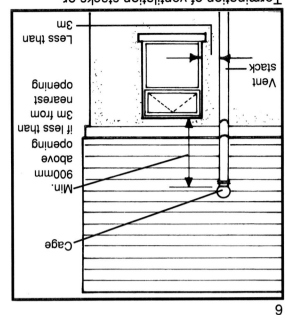

Termination of ventilation stacks or ventilating part of discharge stacks.

Less than 3m

Vent stack

Min. 900mm above opening if less than 3m from nearest opening

Cage

6

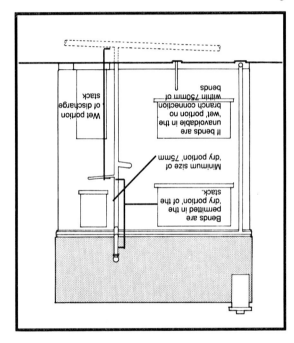

Wet portion of discharge stack

If bends are unavoidable in the 'wet' portion no branch connection within 750mm of bends

Minimum size of 'dry portion' 75mm

Bends are permitted in the 'dry portion' of the stack.

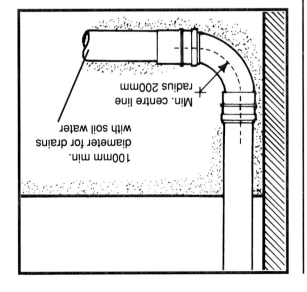

100mm min. diameter for drains with soil water

Min. centre line radius 200mm

MATERIALS FOR PIPES, FITTINGS AND JOINTS

Any of the materials shown in the table opposite may be used. Pipes should be firmly supported without restricting thermal movement. Reference should be made where necessary to the requirements of Part B of the first schedule to the Building Regulations, 1991 and Guidance in Technical Guidance Document B relating to penetration of fire stopping elements and of fire stopping provisions.

Airtightness

The pipes, fittings and joints should be capable of withstanding an air or smoke test of positive pressure of at least 30mm water gauge for at least 3 minutes. During this time every trap should maintain a water seal of at least 25mm. Smoke testing is not recommended for uPVC pipes.

Materials for sanitary pipe work	
Pipe material	Irish/British standard
uPVC	BS 4514
Polypropylene	BS 5245
Plastics	BS 5255, BS 5556
ABS, MUPVC Polyethylene	I.S. 134, I.S. 135
Trap material	
Plastics	BS 3943

FOUL WATER DRAINAGE PIPEWORK UNDERGROUND

Layout

1 The layout of the drainage system should be as simple as possible.

2 Changes in direction and gradient should be minimised and as easy as practicable.

3 Access points should be provided only if blockages cannot be cleared without them.

4 Connections of drains to other drains or to sewers should be made obliquely, and in the direction of flow.

5 The drainage system should be ventilated by a flow of air. A ventilated pipe should be provided at or near the head of each main drain, and to any branch longer than six metres and on a drain fitted with an intercepting trap. Ventilated discharge pipes can be used, see page 299.

6 Pipes should be laid to even gradients, and any change of gradient should be combined with an access point.

7 Pipes should normally be laid in straight lines.

8 Pipes may be laid to a slight curve providing the curve can be cleared of blockages. The curve should be located as follows:

(a) Close to inspection chambers or manholes.

(b) At the foot of discharge and ventilating stacks.

Any curves should have as large a radius as possible.

9 Where drains run under or near buildings, on piles or beams, in common trenches or in unstable ground, special precautions should be taken to accommodate the effects of settlement.

MATERIALS FOR UNDERGROUND PIPES AND JOINTING

Any of the materials shown in the table below.

1 Joints in the pipe should be appropriate to the material of the pipes.

2 To minimise the effects of any differential settlement, pipes should have flexible joints.

3 All joints should remain watertight under working and test conditions.

4 Nothing in the pipes, joints or fittings should project into the pipeline or cause an obstruction.

Any of the materials shown in the table below may be used.

Materials for below ground gravity drainage.	
Material	**Irish/British standard**
Rigid pipes	
Asbestos cement	I.S 243, BS 3656
Vitrified clay	I.S. 106, BS 65
Concrete	I.S. 6, I.S. 166, BS 5911
Flexible pipes	
uPVC	I.S. 424, BS 4660

CLEARANCE OF BLOCKAGES

A sufficient number of access points should be provided for clearing blockages in the drainage system which cannot be reached by any other means. The location, spacing and type of access point will depend on the layout, depth and size of the drainage runs.

The following requirements are for normal methods of rodding (which need not necessarily be in the direction of the flow) but are not for mechanical means of clearing.

1　Access points should be one of four types:

(a) Rodding eyes — capped extensions of the pipes.

(b) Access fittings — small chambers on (or an extension of) the pipes but not with an open channel.

(c) Inspection chambers — shallow chambers e.g., an armstrong–junction, with working space at ground level.

(d) Manholes — large chambers with working space at drain level.

The table opposite shows the depth at which each type of access fitting may be used and the recommended dimensions they should have.

2　Siting of access points — Access points should be provided at the following locations:

(a) On or near the head of each drain run and

(b) At a bend and at a change of gradient and

(c) At a change of pipe size and

(d) At a junction unless each run can be cleared from an access point.

Minimum dimensions for access fittings and chambers		
Type of access point	**Depth to invert (metres)**	**Internal size length x width**
Access fittings (e.g. small chambers) Small Large	0.6 or less	150 x 100 300 x 100
A.J.	0.6 or less	300 x 300
Inspection chamber	1.0 or less	450 x 450
Manhole	2.7 or less over 2.7	1200 x 750 1200 x 840
Shaft	over 2.7	900 x 840

Note:
Cover to access fittings and inspection chambers to be same dimension as the fitting/chamber.
Cover to manhole to be 600 x 600 min.

CLEARANCE OF BLOCKAGES

continued.

3 Provide access points to all long drainage runs — the distance between access points will depend on the type of access used but should not be more than that shown in the table opposite for drains up to and including 300mm diameter.

4 Inspection chambers and manholes should have removable non–ventilating covers of durable material such as:

(a) cast iron
(b) cast or pressed steel
(c) precast concrete
(d) uPVC

and be of suitable strength.

Inspection chambers and manholes within buildings should have mechanically fixed airtight covers unless the drain itself has water–tight access covers. Manholes deeper than one metre should have metal step irons or fixed ladders.

Maximum spacing of access points in metres

From	to	Junction	Inspection Chamber	Manhole
Start of external drain			22	45
Rodding eye		22	45	45
Access fitting Small 150 dia. 150x100	12		22	22
Large 300x100	22		45	45
Inspection chamber, A. J.	22		45	45
Manhole	45		45	90

5 Construction of access points.

Materials for access points

Materials	Irish/British standard
Inspection chambers and manholes — Bricks and Blocks	I.S. 20, I.S. 91. I.S. 189.
Vitrified clay bricks	BS 65
Precast concrete	I.S. 6, I.S. 166, BS 5911
In–situ concrete	I.S. 325, I.S. 326
Plastics	IAB certificates (Irish Agrément Board)
Rodding eye and access fittings (excluding frames and cover)	The same as below ground drainage materials, see table on page 302.

The construction of access points should be such that they contain the foul water under working and test conditions and they resist the entry of rainwater/ground water.

CLEARANCE OF BLOCKAGES
continued.

6

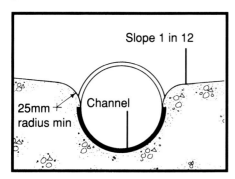

Channels and branches should be benched up at least to the top of the outgoing pipe and at a slope of 1 in 12. The benching should be rounded at the channel with a radius of at least 25mm.

7   Where half round channels are used in inspection chambers and manholes, the branches should discharge into the channel at or above the level of the horizontal diameter.

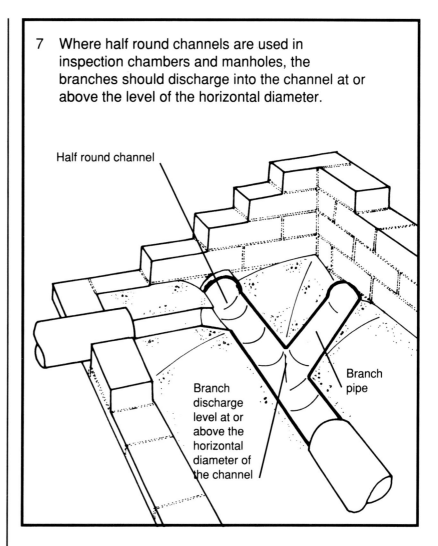

8   Where the angle of the branch is more than 45° a three quarter section branch should be used.

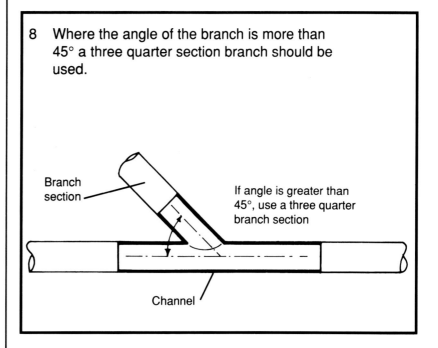

TRENCH EXCAVATION

1   Trenching should not be carried out too long in advance of pipelaying, and backfilling should take place as soon as possible.

2   The trench should be kept as narrow as practicable, but must allow adequate room for jointing and placing and compacting the backfill. Trenches should be excavated with vertical sides to a height of at least 300mm above the top of the pipe, where flexible pipes are being laid.

3   Excavated spoil should be kept at least 0.5m back from the edge of the trench, and all loose stones should be removed from the heap.

4   The trench bottom should be at a level which will accommodate laying the pipes on the prepared underbed. The bottom should be examined for the presence of soft spots or hard objects which should be removed and filled with well tamped bedding materials.

5   Where a delay in pipelaying is envisaged, the bottom layer of 300mm should not be removed until immediately before laying commences.

6   All excavations deeper than 1.25m should be sloped back to a safe angle.

PIPE GRADIENTS AND SIZES.

1   Drains should be laid to falls and have sufficent capacity to carry the flows.

2   The minimum allowable diameters and gradients are illustrated in the following diagrams.

A   Small single dwelling

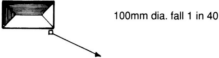

100mm dia. fall 1 in 40

B   Single dwelling discharging to a septic tank or main sewer

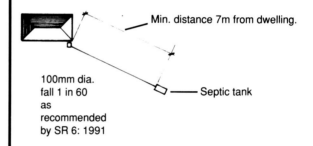

Min. distance 7m from dwelling.

100mm dia. fall 1 in 60 as recommended by SR 6: 1991

Septic tank

C   2 dwellings

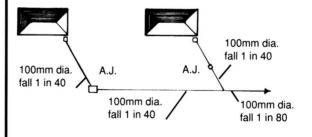

100mm dia. fall 1 in 40

A.J.

A.J.

100mm dia. fall 1 in 40

100mm dia. fall 1 in 40

100mm dia. fall 1 in 80

D   3 – 8 dwellings

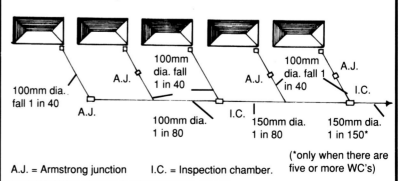

100mm dia. fall 1 in 40

A.J.

A.J.

100mm dia. fall 1 in 40

100mm dia. 1 in 80

I.C.

100mm dia. fall 1 in 40

A.J.

150mm dia. 1 in 80

A.J.

I.C.

150mm dia. 1 in 150*

(*only when there are five or more WC's)

A.J. = Armstrong junction   I.C. = Inspection chamber.

PIPELAYING AND JOINTING

1 The pipes should be joined in the trench and laid on the prepared bed so that they maintain substantially continuous contact with the bed. Small depressions should be left in the bed to accommodate the joints.

Levelling devices such as bricks or pegs should be removed, and any resulting voids should be filled before backfilling.

2 Particular care should be taken when installing uPVC pipes at temperatures below 3°C. When the temperature of uPVC pipe is below 0°C pipelaying should not be carried out.

3 The types of bedding and backfilling for rigid pipes of standard strength laid in a trench of any width are shown in the table and diagrams below.

Limits of cover in metres for standard strength rigid pipes in any width of trench							
Pipe bore	Bedding class	Fields & Gardens		Light traffic roads		Heavy traffic Roads	
		Min.	Max.	Min.	Max.	Min.	Max.
100	D or N	0.4	4.2	0.7	4.1	0.7	3.7
	F	0.3	5.8	0.5	5.8	0.5	5.5
	B	0.3	7.4	0.4	7.4	0.4	7.2
150	D or N	0.6	2.7	1.1	2.5	—	—
	F	0.6	3.9	0.7	3.8	0.7	3.3
	B	0.6	5.0	0.6	5.0	0.6	4.6

Key to diagram opposite

Selected fill free from stones larger than 40mm, lumps of clay over 100mm, timber, frozen material and vegetable matter.

Granular material — 10mm single size or 5mm – 14mm graded aggregate, both with minimum, fines content.

Note:

1 Provision may be required to prevent ground water flow in trenches with class N, F or B type bedding.

2 Where there are sockets these should be not less than 50mm above the floor of the trench.

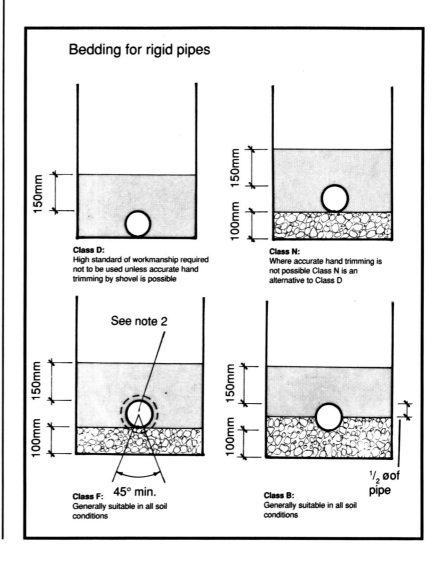

Bedding for rigid pipes

Class D:
High standard of workmanship required not to be used unless accurate hand trimming by shovel is possible

Class N:
Where accurate hand trimming is not possible Class N is an alternative to Class D

See note 2

Class F: 45° min.
Generally suitable in all soil conditions

Class B:
Generally suitable in all soil conditions

½ ø of pipe

PIPELAYING AND JOINTING
continued.

Key to diagram opposite.

 Selected fill free from stones larger than 40mm, lumps of clay over 100mm, timber, frozen material and vegetable matter.

 Selected fill or granular fill free from stones larger than 40mm.

 Granular material — 10mm single size or 5mm – 14mm graded aggregate, both with minimum, fines content.

5 When the pipes have been laid, bedded and surrounded with granular material, the remainder of the backfill may be material excavated from the trench, provided it is free from heavy stones or other objects. It should be compacted in layers.

4 Flexible pipes will become deformed under load and require support to limit the deformation to 5 per cent of the diameter of the pipe. The bedding and backfilling should be as shown below.

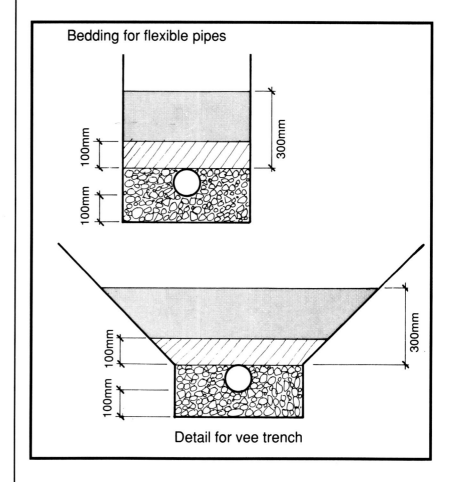

Bedding for flexible pipes

Detail for vee trench

PIPELAYING AND JOINTING
continued.

Depth of pipe cover
The depth of cover will depend on the levels of the connections to the system. Where connections leave gullys, inspection chambers, access junctions etc., the depth of cover will be determined by the depth of the outlet from the fitting. Main house drains should not have a depth of cover of less than the following:

0.9m under carriageways.

0.6m under agricultural land, or gardens within the curtilage of dwellings.

Special Protection
1) Where minimum adequate cover for rigid pipes cannot be achieved the pipes should be protected by concrete encasement not less than 100mm thick and having movement joints formed with compressible board at each socket or sleeve joint face.

2 Where flexible pipes are not under a road and have less than 0.6m cover, they should have concrete paving slabs laid as bridging above the pipes with at least 75mm of granular material between the top of the pipe and the underside of the slabs.

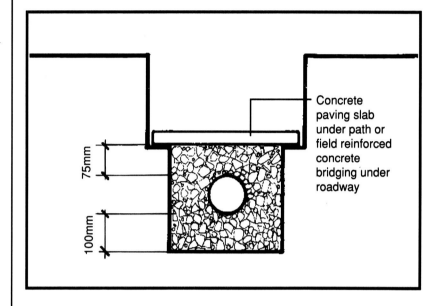

Concrete paving slab under path or field reinforced concrete bridging under roadway

3 Where flexible pipes are under a road and have less than 0.9m cover, reinforced concrete bridging or a reinforced concrete surround should be used instead of paving slabs.

TESTING

After laying and before backfilling drains and sewers should be subjected to either of the following:

1) Water test, or

2) Air test.

Water test: Drains and sewers up to 300mm diameter should be capable of withstanding a water test to a pressure equal to 1.5m head of water pressure above the invert at the head of the drain.

Where the drain is water tested using a stand pipe of the same diameter as the drain, the section of drain should be filled and left to stand for 2 hours and topped up. The leakage over 30 minutes should then be measured and should not be more than 0.05 litres for each metre run of drain for a 100mm drain — a drop in water level of 6.4mm/m, and not more than 0.08 litres for a 150mm drain — a drop in water level of 4.5mm/m.

Any leakage, which causes a drop in the test water level, should be investigated and the defective part of the work removed and made good.

Air test: It is sometimes more convenient to test sewers by means of internal air pressure. However while an excessive drop in pressure when employing the air test may indicate a defective line, the location of the leakage may be difficult to detect and the leakage rate cannot be measured. Air pressure is also affected by temperature changes. Consequently, failure to pass this test is not necessarily conclusive and when failure does occur, a water test as previously described should be carried out, and the leakage rate determined before a decision as to acceptance or rejection is made. Failure to pass an air test is very often attributable to faults in the plugs or testing apparatus.

The air test is carried out after laying, including any necessary concrete or other haunching or surrounding and backfilling. Gravity drains and private sewers up to 300mm should be capable of withstanding an air test ensuring that the head of water on a manometer does not fall by more than 25mm in a period of 5 minutes for a 100mm water gauge test pressure and 12mm for a 50mm water gauge test pressure.

Advice and guidance on test procedures and results is available from the pipe manufactures.

SEPTIC TANKS

Guidance on septic tanks is contained in Standard Recommendation 6 (SR6) : 1991, Septic Tank Systems : Recommendations for domestic effluent treatment and disposal from a single dwelling house.

This is the second edition of these recommendations to be published and there are a number of departures from the now superseded SR6 : 1975. Some of the more important of these departures include :

◆ More severe site test requirements.

◆ Site assessment is introduced.

◆ Introduction of minimum distance between percolation areas and ground water sources.

◆ Advice is given on site improvements in the event of test failure.

◆ Minimum capacity of septic tanks for different dwelling house sizes.

As a result of these changes and the additional requirements, it is essential to be familiar with the current SR6 : 1991 to ensure proper construction, layout and maintenance of septic tanks.

Standard Recommendation 6 is published by EOLAS and is available from the National Standards Authority of Ireland (NSAI), Glasnevin, Dublin 9.

Note: Where sites are unsuitable for septic tanks, systems based on biological filtration, activated sludge or other approved systems e.g. Ágrement Certification, may be appropriate, subject to Local Authority approval.

SOAKAWAYS

Soakaways have been the traditional way to dispose of stormwater from buildings and paved areas remote from a public sewer or watercourse. Soakaways are seen increasingly as a more widely applicable option alongside other means of stormwater control and disposal.

Soakaways must store the immediate stormwater run–off and allow for its efficient infiltration into the adjacent soil. They must discharge their stored water sufficiently quickly to provide the necessary capacity to receive run–off from a subsequent storm. The time taken for discharge depends upon the soakaway shape and size, and the surrounding soil's infiltration characteristics. They can be constructed in many different forms and from a range of materials.

Shape and size.

Traditionally, soakaways for areas less than 100m^2 have been built as square or circular pits; they are either filled with rubble or lined with dry–jointed brickwork or pre–cast perforated concrete ring units surrounded by suitable granular backfill. Code of Practice for Building Drainage, BS 8301, suggests that soakaways may take the form of trenches that follow convenient contours. Trenches have large internal surface areas for infiltration of stormwater for a given stored volume compared with square or circular shapes.

Soakaways design.

Reference should be made to the detailed design guidance for soakaways as outlined in BRE Digest 365, Soakaway Design.

General consideration.

Soakaways should not normally be constructed closer than 5m to building foundations. In soil and fill material subject to modification or instability, the advice of a specialist geotechnologist should be sought as to the advisability and siting of a soakaway. Care should be taken in limestone areas.

Note:
No general guidance can be given on the size of soakaways. Size depends on the volume of water discharging to the soakaway and the geology of the site.

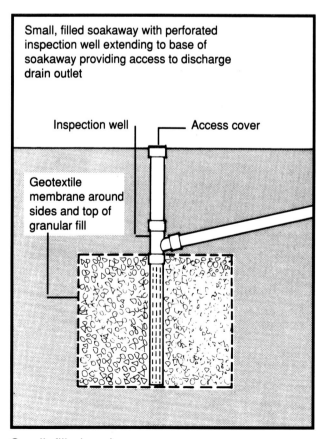

Small, filled soakaway with perforated inspection well extending to base of soakaway providing access to discharge drain outlet

Inspection well   Access cover

Geotextile membrane around sides and top of granular fill

Small, filled soakaway

FIRE

INTRODUCTION

Part B of the Building Regulations sets out general requirements in respect of fire safety in all buildings, including housing. Technical Guidance Document B, published by the Department of the Environment, gives detailed guidance on how these requirements can be successfully complied with. In addition, Part J of the Regulations sets out requirements relating to heat producing appliances such as central heating boilers, and Technical Guidance Document J gives guidance on compliance with those requirements. On the following pages is set out a brief summary of the principal recommendations of Technical Guidance Documents B and J as they apply to housing. For further details on this topic the reader is directed to the Technical Guidance Documents themselves and to the HomeBond publication "Housing and the Building Regulations".

MEANS OF ESCAPE

In conventional single–storey and two–storey housing, the internal layout will usually provide an acceptable availability of escape routes to occupants. Certain conditions may however constitute risks that require additional alternatives to be provided. Examples of this would include what are called "inner room" conditions (where one room is entered through another), and in houses of an "open–plan" design where the stairs to first floor accommodation rises directly from a room at ground floor level, rather than from an enclosed hallway.

In these situations, alternative means of escape should be provided from the inner room or first floor rooms, as the case may be, via suitably located and sited "escape windows" (or doors). Escape windows should incorporate an opening section of at least 850mm high by 500mm wide the bottom of which should be between 600mm and 1100mm above floor level and should be located above an area of ground which is:

(a) free of permanent obstructions

(b) suitable for the use of a ladder to provide egress from the window, and

(c) readily accessible for ladder access to the windows by fire brigade personnel.

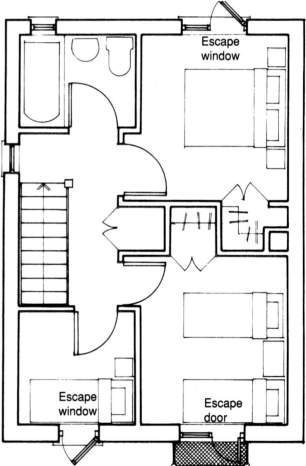

Ground floor
Typical two storey floor plan, with open plan ground floor layout.

First floor
Provide escape windows/doors at first floor level where ground floor is open plan, as illustrated opposite.

Note:
Where escape doors are provided at upper levels they should open onto a balcony. Any such balcony should be provided with a protective barrier in accordance with BS 6180.

MEANS OF ESCAPE
continued.

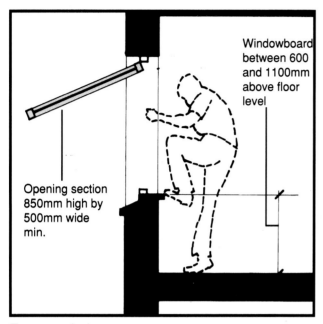

Windowboard between 600 and 1100mm above floor level

Opening section 850mm high by 500mm wide min.

Escape window

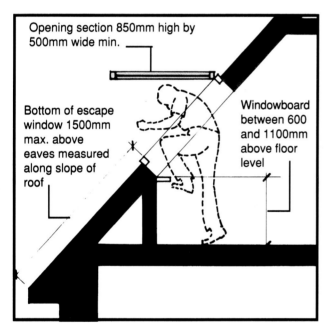

Opening section 850mm high by 500mm wide min.

Bottom of escape window 1500mm max. above eaves measured along slope of roof

Windowboard between 600 and 1100mm above floor level

Escape window in attic room.

Where the stairway does not connect to an external door via a circulation space, two independent escape routes should be provided to the outside from the bottom of the stairs. If this is not done, escape windows/doors are required.

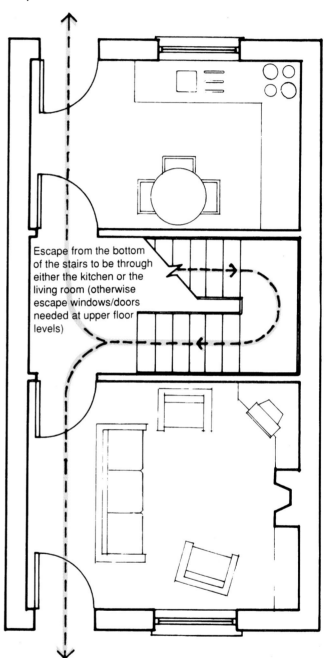

Escape from the bottom of the stairs to be through either the kitchen or the living room (otherwise escape windows/doors needed at upper floor levels)

Floor plan incorporating internal stairway.

MEANS OF ESCAPE
continued.

In houses where the design incorporates three storeys of accommodation, additional protection of the hall/stairs/landings areas is considered necessary.

This protection includes the provision of self-closing fire resisting doors (FD 20) and half-hour fire-resisting partitions between all rooms and the hall/stairs/landing areas.

It should also be noted that the floors in houses of this configuration are required to have full half-hour fire resistance (not modified half-hour fire resistance). See page 320 for details of floor construction with half hour fire resistance.

Illustrated on this page are floor plans of a typical three storey house highlighting the protection of the hall/stairs/landing areas and the location of fire doors.

Where a fire door is used in circumstances such as those illustrated here, care should be taken to ensure that the door as installed complies with the specification as set out in the fire test certificate for the door in question.

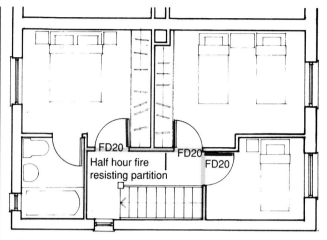

Second floor

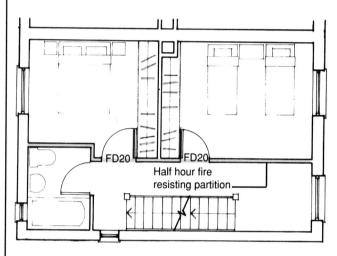

First floor

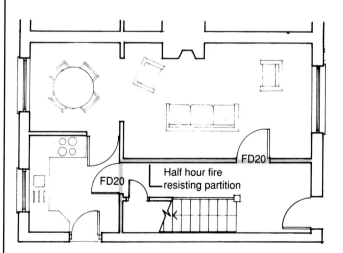

Ground floor

MEANS OF ESCAPE
continued.

Another precaution considered necessary as a means of ensuring early escape in case of the outbreak of fire is the installation of smoke detectors. These detectors should be installed in all houses in accordance with the following guidelines:

◆ Detectors should be smoke alarms complying with BS 5446: Part 1. It should be noted that Technical Guidance Document B does not exclude the use of battery-operated detectors. A subsequent revision of the relevant British Standard recommends mains-operated alarms. This provision may be taken into account in future amendments to the Technical Guidance Document.

◆ Number and location of smoke detectors will depend on the type and layout of the house in question. Detailed guidance on this aspect of their installation is given in appendix D of BS 5588: Part 1:1990. In bungalows, for example, the corridor between living and sleeping accommodation is a suitable location. In a typical two– storey house, the best place for a single smoke alarm is in the hallway above the bottom of the staircase.

◆ Locate detectors on ceilings at least 300mm from walls and light fittings, preferably in a central position. Detectors designed for wall mounting should be fitted between 150 and 300mm below ceiling level.

◆ Locate detectors where they are readily accessible for testing and maintenance.

◆ Don't locate detectors in rooms which tend to get very hot or very cold.

◆ Don't locate detectors in bath/shower rooms, kitchens or garages where fumes may trigger the alarm. Don't locate detectors above heaters or radiators.

◆ Don't locate detectors on surfaces likely to be much warmer or colder than the rest of the space.

As stated opposite, the type and layout of the house will influence the appropriate choice of detection/alarm system. For example, in large houses such as country mansions or those which have a complex layout it may be necessary to incorporate additional detectors or a more elaborate installation powered from mains electrical supply. Further guidance on the installation of smoke detectors is given in BS 5588: Part 1 – Fire Precautions in the Design and Construction of Buildings: Code of Practice for Residential Buildings.

WALL AND CEILING LININGS

Wall and ceiling linings can contribute significantly to the spread of fire and are therefore controlled. Plastered surfaces are acceptable but some other materials are restricted in their use. Timber sheeting used as a lining may only be used in small panels (of 5.0m^2 or less). Such panels should be separated by at least 2m from any other such panel and the total area of such panels in any room should not exceed 20m^2 or half the floor area of the room, whichever is less. This relaxation for untreated panels of limited size and disposition applies only to wall linings – it does not apply to ceiling linings.

Alternatively sheeting may be treated by an appropriate method, such as a fire retardant varnish or paint, to give it a "Class 1" Surface Spread of Flame rating. It should also be noted that the use of timber sheeting applied directly to the underside of floor joists will generally not meet the recommended level of fire resistance. In such instances the ceiling should be formed by plasterboard in the usual way prior to the fixing of the timber sheeting. The requirements in respect of Class 1 surface spread of flame apply everywhere in houses apart from bathrooms, see sketch opposite.

The use of polystyrene tiles as a ceiling finish is not permitted.

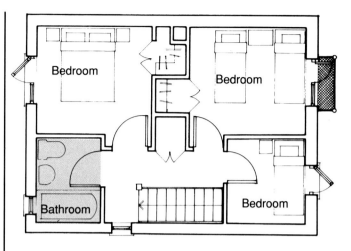

First floor

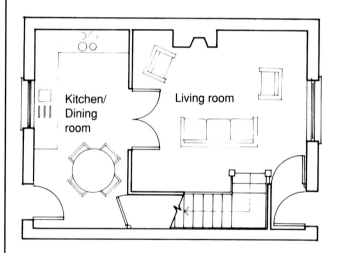

Ground floor

Typical two story house plan – the requirements for internal wall and ceiling linings are Class 0 or 1 to all rooms except bathrooms or shower rooms which can be Class 0, 1, 2 or 3, which are measures of the rate of flame spread on the surfaces of materials.

FIRE RESISTANCE

The nature of conventional house construction is such that the standard requirements for fire resistance for such elements as floors and walls are usually automatically met by such construction. However, there are certain situations where particular attention is required to ensure that an adequate level of resistance to fire is incorporated into the construction of houses. These are as follows:

◆ Floors in three–storey houses and over garages should have full half–hour fire resistance, not modified half–hour fire resistance which is normally acceptable in two–storey housing. Attention may be required to the detailed specification of ceiling and floor finishes to ensure that these requirements are met. For example, a timber floor incorporating a ceiling consisting of 12.5mm plasterboard with all joints taped and filled and backed by timber is deemed to have modified half–hour fire resistance. Such a construction is acceptable in two–storey construction but not in three–storey construction or between a garage and other accommodation. The table opposite sets out a range of common options for the construction of floors with full half–hour fire resistance.

Note: Other means of achieving modified half hour fire resistance can be found in 'Guidelines for the construction of fire resisting structural elements' a report by the Building Research Establishment.

FLOORS OF 1/2 HOUR FIRE RESISTANCE

Ceiling specification	Floor type (explained below)
one layer of 12.5mm plasterboard with joints taped and filled and backed by timber	3
two layers of plasterboard with joints staggered and joints in outer layer taped and filled:	
25mm total thickness	1
22mm total thickness	2*
19mm total thickness	3*
one layer of 12.5mm proprietary fire grade plasterboard with joints taped and filled	1,2,3
one layer of 9.5mm plasterboard finished with 10mm min. light–weight aggregate gypsum plaster	2*, 3
one layer of 12.5mm plasterboard finished with:	
13mm light weight aggregate gypsum plaster	1
10mm lightweight aggregate gypsum plaster	2, 3
5mm gypsum board finish plaster	2, 3

*supports not exceeding 450mm centres

Floor types referred to above are as follows:

1. Any structurally suitable flooring of timber or particle boards on timber joists not less than 37mm wide.

2. Timber floor boarding, plywood or wood chipboard tongued and grooved and not less than 15mm (finished) thickness on timber joists not less than 37mm wide.

3. Timber floor boarding, plywood or wood chipboard, tongued and grooved and not less than 21mm (finished) thickness on timber joists not less than 37mm wide.

FIRE RESISTANCE
continued.

As stated earlier in three storey houses additional protection of the hall/stairs and landings is considered necessary. As well as providing fire resisting doors, half hour fire resisting partitions are required between all rooms and the hall/stairs and landing area.

Depending on whether the partition is loadbearing or non–loadbearing, to achieve half hour fire resistance partitions should be constructed in accordance with the tables opposite. Blockwork partitions will also usually satisfy the requirements for half–hour fire resistance.

Note:
Reference should also be made to pages 123 to 126 which deal with the construction of loadbearing and non–loadbearing timber stud partitions.

The table on these pages have been derived from 'Guidelines from the construction of fire resisting elements', a report by the Building Research Establishment.

FIRE RESISTANT LOADBEARING INTERNAL TIMBER STUD PARTITIONS
Nature of construction and materials to achieve half hour fire resistance
44mm min. timber studs at 600mm centres (except where indicated) faced on each side with:
One layer of 12.5mm plasterboard with all joints taped and filled.
One layer of 9.5mm* plasterboard with a finish of lightweight aggregate gypsum plaster – thickness of plaster 10mm.
* Supports not to exceed 450mm centres.

FIRE RESISTANT NON–LOADBEARING INTERNAL TIMBER STUD PARTITIONS
Nature of construction and materials to achieve half hour fire resistance
Timber studs at 600mm centres (except where indicated) faced on each side with:
One layer of 12.5mm plasterboard with all joints taped and filled.
Two layers of plasterboard with joints staggered, joints in outer layer taped and filled – total thickness for each face 19mm*.
9.5mm* thick plasterboard with 10mm min lightweight gypsum plaster finish.
*Supports not to exceed 450mm centres.

FIRE RESISTANCE
continued.

Doors forming direct connections between garages and areas within houses should be self–closing type FD 20, and should open over a 100mm upstand or step down from the house to the garage. In addition, any garage attached to a house must be separated from the rest of the accommodation by walls, floors or ceilings having full half–hour fire resistance.

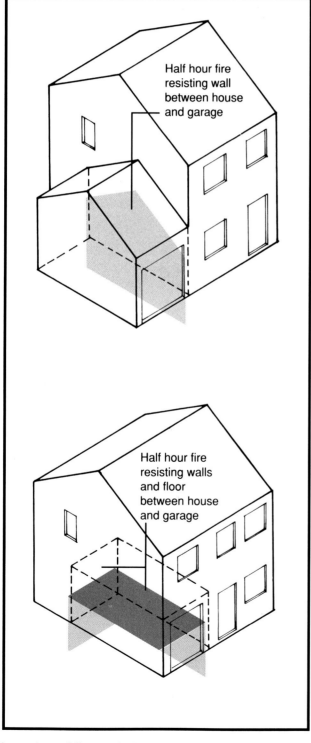

Half hour fire resisting wall between house and garage

Half hour fire resisting walls and floor between house and garage

Location of fire resisting walls, floors and ceilings

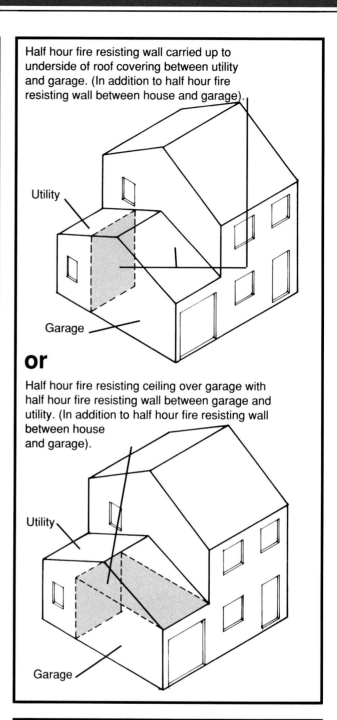

Half hour fire resisting wall carried up to underside of roof covering between utility and garage. (In addition to half hour fire resisting wall between house and garage).

Utility

Garage

or

Half hour fire resisting ceiling over garage with half hour fire resisting wall between garage and utility. (In addition to half hour fire resisting wall between house and garage).

Utility

Garage

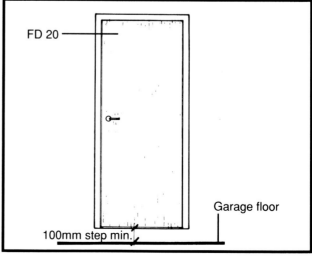

FD 20

Garage floor

100mm step min.

Fire door separating garage from rest of house

FIRE RESISTANCE
continued.

Half–hour fire resisting ceiling with no floor above

In situations such as illustrated here, it may be necessary to provide half hour fire resistance in a ceiling which does not incorporate floor boarding on its upper surface. Set out on this page are details of the construction of such ceilings.

Construction specification: Min. 44mm wide joists with 12.5mm proprietary fire grade plasterboard, all joints taped and filled and backed by timber, with minimum 60mm mineral wool mat between joists. (Note that the thermal insulation requirements of the Building Regulations may require a greater thickness of material than that deemed necessary to satisfy fire safety requirements).

Note:
In prefabricated truss rafter construction where the width of ceiling rafter is usually 34 or 41mm and thus lower than the 44mm required it is acceptable to screw fix (with 50mm long wood screws at 300mm max. centres) a min. 50 x 25mm wide batten to the rafter of the truss to increase the overall width of the rafter as illustrated opposite.

Note:
As an alternative to this detail the wall between the garage and house can be brought up to the underside of the roof covering in half hour fire resisting construction.

Note:
A fire door separating a house from a garage should be a door certified by test as being an FD20 door, and should be installed in compliance with the detailed specification as set out in the fire test certificate for the door type in question.

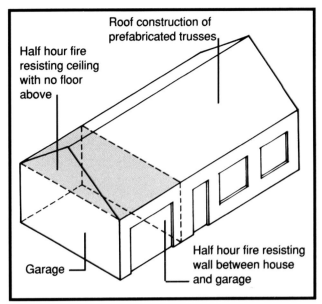

Typical bungalow incorporating garage.
A location where a half hour fire resisting ceiling with no floor above may occur.

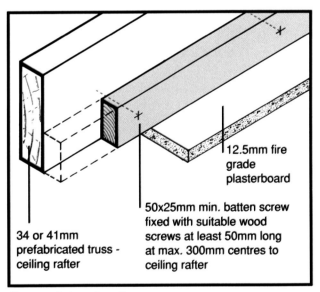

Batten fixing detail.

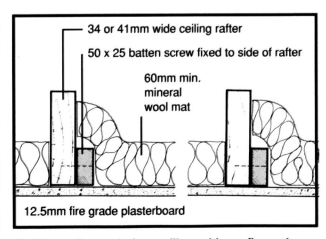

Half hour fire resisting ceiling with no floor above

FIRE RESISTANCE
continued.

Fire–stopping of party walls is an essential part of ensuring that fire does not readily spread from house to house. See pages 140 and 141 of this Manual for guidance in respect of good practice in this area, in particular at the junction of the party wall with the roof and at eaves level where it is important to fire–stop the space contained by the fascia and soffit. The practice illustrated on page 292 to ensure adequate sound insulation will also ensure adequate fire resistance by avoiding timber floor joists being built into or through party walls.

In semi–detached and terraced houses it is not required to provide fire stopping in the form of a vertical cavity barrier at the junction of party wall and external wall, **provided** that the cavity at door and window head is closed. This can be achieved using either of the methods illustrated on this page.

Note:
It is not necessary to close the cavity at door and window heads in a detached house. (However, it is likely that conventional construction practice will normally ensure that the cavity is closed at the head)

◆   In all situations the cavity must be closed at the top of the wall.

◆   The illustrations in this publication show the provisions considered necessary for the semi-detached and terraced situation.

Technical Guidance Document B of The Building Regulations does not make any specific requirements for closing at jambs.

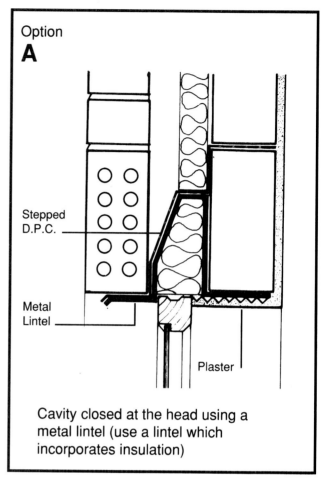

Option **A**

Stepped D.P.C.

Metal Lintel

Plaster

Cavity closed at the head using a metal lintel (use a lintel which incorporates insulation)

OR

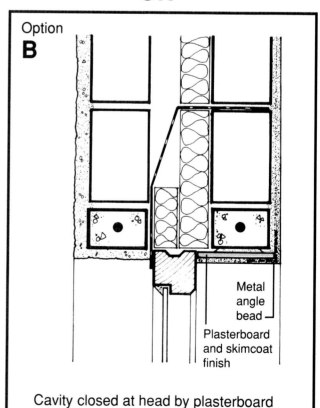

Option **B**

Metal angle bead

Plasterboard and skimcoat finish

Cavity closed at head by plasterboard fixed by dabs to underside of lintel and tight against window frame with skim plaster finish

FIRE RESISTANCE
continued.

◆ In certain circumstances, steel members such as RSJ's and Universal Beams will be incorporated into the structure of a house. Where such steel is giving structural support to floors or walls the steel should have minimum half–hour fire resistance – this will usually require that some form of fire protection medium is applied to the steel. Typical protection media include plasterboard, proprietary fire protection boards and intumescent paints. Precise specifications vary according to the level of fire resistance required in specific instances, and the weight and cross–sectional characteristics of the steel member. Suppliers/manufacturers of fire protection media provide detailed guidance on the use of their products.

ROOF COVERINGS

Roof coverings are required to prevent ready penetration of fire and also to limit flame spread on their surfaces. This is achieved by requiring materials to have specified roof "designations". The common roofing materials, such as concrete tiles and fibre cement slates, used in housebuilding readily meet these requirements. Flat roofing materials may need additional protection to achieve compliance. For example, bituminous felt requires to be covered by a finish such as bitumen–bedded chippings to be deemed acceptable as a roof finish close to boundaries. Where such materials are being used, the manufacturer's/supplier's guidance should be sought in respect of any surface treatments necessary to achieve the recommended designation.

FIRE FIGHTING ACCESS

Housing estates should be laid out to ensure ready access for fire tenders in the event of a fire. This will require that roads and access generally is adequate for fire tenders. The following Table summarises the requirements of the TGD in this regard.

The layout of most, if not all, estates to accommodate everyday vehicular traffic will automatically ensure that these requirements are met.

Vehicle access route specifications	
Appliance type	Access specification
Pump	Minimum width of roadway between kerbs 3.7m.
Pump	Minimum width of gateway between kerbs 3.1m.
Pump	Minimum turning circle between kerbs 16.8m.
Pump	Minimum turning circle between walls 19.2m.
Pump	Minimum clearance height 3.7m
Pump	Minimum carrying capacity 12.5 tonnes.

RADIATION ONTO BOUNDARIES

Large openings in external walls, such as doors and windows, may in the event of a fire give rise to radiation which may in turn endanger adjoining buildings and their occupants. To control this risk, the Technical Guidance Document limits the extent of such "unprotected areas" depending on their distance to boundaries. The accompanying sketch illustrates the restrictions that apply on openings within one metre of the boundary.

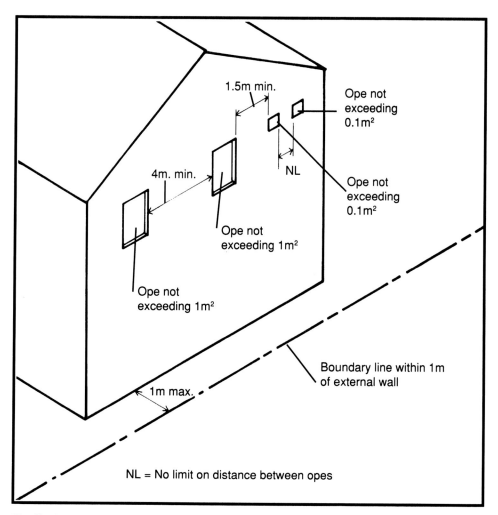

1.5m min.

Ope not exceeding 0.1m²

4m. min.

NL

Ope not exceeding 0.1m²

Ope not exceeding 1m²

Ope not exceeding 1m²

Boundary line within 1m of external wall

1m max.

NL = No limit on distance between opes

Radiation onto boundaries; unprotected areas less than 1m from the boundary line – only unprotected areas of the size and disposition illustrated here should be used.

RADIATION ONTO BOUNDARIES
continued.

For openings a metre or more from the boundary the limit of unprotected areas can be calculated using the Table set out here – a typical example is given to illustrate its application.

Permitted unprotected areas in dwellings not greater than 10m in height.	
Maximum total percent of unprotected areas (%)	Minimum distance between side of building and relevant boundary (m)
8	1
20	2.5
40	5.0
60	7.5
80	10.0
100	12.5

Note: Intermediate values may be obtained by interpolation.

Example

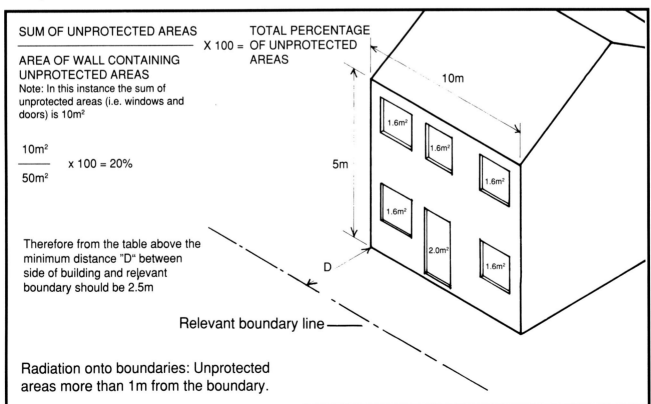

$$\frac{\text{SUM OF UNPROTECTED AREAS}}{\text{AREA OF WALL CONTAINING UNPROTECTED AREAS}} \times 100 = \text{TOTAL PERCENTAGE OF UNPROTECTED AREAS}$$

Note: In this instance the sum of unprotected areas (i.e. windows and doors) is 10m²

$$\frac{10m^2}{50m^2} \times 100 = 20\%$$

Therefore from the table above the minimum distance "D" between side of building and relevant boundary should be 2.5m

Relevant boundary line ——

Radiation onto boundaries: Unprotected areas more than 1m from the boundary.

RADIATION ONTO BOUNDARIES
continued.

For the purpose of assessing radiation risks the boundaries in typical housing layout will be as follows.

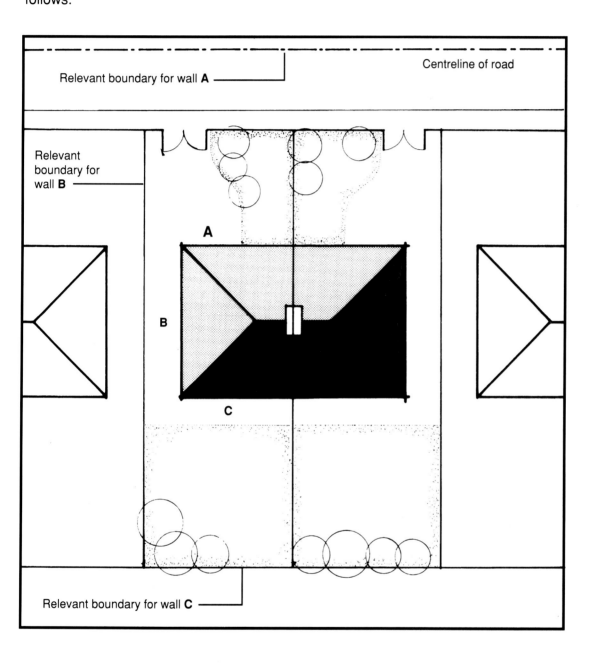

ROOFLIGHTS

Technical Guidance Document B of the Building Regulations, gives recommendations on the use of glass and plastic rooflights. The principal provisions relevant to housing are summarised below.

Glass rooflights

Appendix E (glazing) gives general guidance on appropriate thicknesses for glass panes. However, notwithstanding that guidance, any glass in a rooflight within 6m of a boundary should be at least 4mm thick unless it is in a rooflight over:

(a) a balcony, verandah, open carport or detached swimming pool, or,

(b) any garage, conservatory or outbuilding whose floor area does not exceed 40m^2.

Plastic rooflights

Where translucent plastic materials are used as rooflights they should be type TP(a), TP(b) or TP(c) as defined in Technical Guidance Document B. Manufacturers and suppliers should be consulted if any doubt exists as to the classification of such materials when used in rooflights

Single skin plastic rooflights

When materials classified as TP(a), TP(b) or TP(c) are used in single–skin rooflights, the rooflights should generally be located at least 6m from any boundary. However in the case of TP(a) or TP(b) materials being used in rooflights to a balcony, verandah, carport or covered way with at least one longer side of the structure open to the air, there is no restriction on the proximity of the rooflight to the boundary. This relaxation also applies to detached swimming pools and to garages, conservatories and outbuildings whose floor area does not exceed 40m^2.

Where a TP(c) material is used in single–skin rooflights in a room or circulation space within a house the area and spacing of such rooflights is limited as illustrated here.

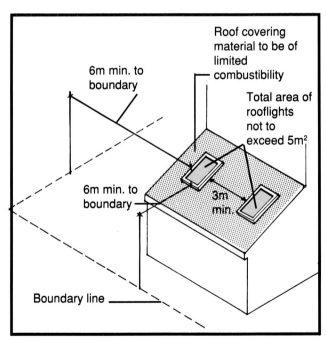

Rooflights of TP(c) material in room/ circulation areas.

ROOFLIGHTS
continued.

Twin–skin rooflights

Where a twin–skin rooflight is made up of two sheets of the same material (i.e. TP(a) inner and outer, TP(b) inner and outer and TP(c) inner and outer) the recommendations in the preceding paragraphs for single–skin rooflights apply.

Set out opposite is a summary of the recommendations where different classifications of materials are used for the inner and outer skins.

Inner skin – TP(a); outer skin – TP(b):
As for single–skin TP(a) or TP(b) as described previously.

Inner skin – TP(a); outer skin – TP(c):
No restriction.

Inner skin – TP(b); outer skin – TP(a):
As for single–skin TP(a) or TP(b) as described previously.

Inner skin – TP(b); outer skin – TP(c):
No restriction.

Inner skin – TP(c); outer skin – TP(a):
No restriction.

Inner skin – TP(c); outer skin – TP(b):
No restriction.

HEATING APPLIANCES

Points to note

The term "room sealed appliance" means an appliance which does not rely on the room in which it is located as a source of air supply e.g. balanced flue appliances. Any appliance in a bath or shower room or in a private garage must be of the room sealed type.

If an appliance is not a room–sealed appliance, the room or space in which it is contained must have a permanent ventilation opening. This ventilation opening should preferably be in an external wall and should be at least the size set out in the table on the opposite page.

Where an appliance is located in such a way that the room or space cannot be ventilated directly to the open air, the permanent ventilation opening may be to an adjoining room or space, provided that the adjoining room or space has a permanent ventilation opening of at least the same size direct to external air.

Note:
It is a requirement of the Department of the Environment Housing Grants Section that any room sealed appliance located in a garage should be:

◆ manufactured for use as a "room-sealed' appliance and installed in strict compliance with the manufacturers instructions.

◆ together with all connections thereto, protected from impact damage either by the provision of suitable barriers or by being located in a position where impact cannot take place.

◆ together with its flue pipe, adequately separated from combustible material.

◆ fitted with a flue terminal so positioned that flue gases and other products of combustion are prevented from entering the building.

◆ fitted with a flue pipe and terminal so positioned and/or shielded as to minimise the risk of danger to persons through contact.

Room sealed appliance – typical gas burning balanced flue boiler.

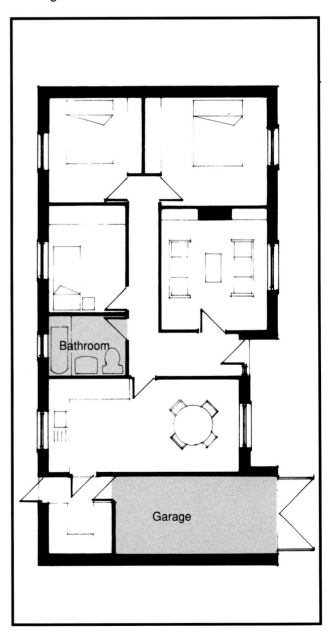

Any appliance in a bath or shower room or in a private garage must be of the room sealed type.

HEATING APPLIANCES
continued.

Air supply to appliances

Type of appliance	Types of ventilation
Solid fuel burning open appliance	An air entry opening or openings with total free area of at least 50% of the appliance throat opening area – as defined in BS 8303: 1986 Code of practice for installation of domestic heating and cooking appliances burning solid mineral fuels.
Other solid fuel appliance	An air entry opening or openings with a total free area of at least 550mm^2 per kW of rated output above 5kW. Where a flue draught stabiliser is used, the total free area should be increased by 300mm^2 for each kW of rated output.
Gas–burning appliances	Permanent ventilation opening of at least 450mm^2 for each kW of appliance input exceeding 7kW.
Gas cooker	The room or space should have an openable window or other means of providing ventilation. If the room or space has a volume of less than 10m^3, a permanent ventilation opening of at least 5000mm^2 should be provided.
Oil–fired appliance	Any room or space containing an oil–fired appliance (other than a balanced flue appliance) should have a permanent ventilation opening of at least 550mm^2 for each kW of rated output above 5kW.

FLUE PIPES

For solid fuel appliances, including open fires, and for oil burning appliances where the flue gas temperature is likely to exceed 260°C, flue pipes should only be used to connect appliances to chimneys and should not pass through roof spaces.

Flue pipes should be manufactured from materials as set out in the table below.

Materials for flue pipes

A	B
Flue pipes to solid fuel appliances and oil–burning appliances whose flue gas temperature is **likely** to exceed 260°C .	Flue pipes to gas fired appliances and oil–burning appliances whose flue gas temperature is **not likely** to exceed 260°C.
◆ Cast iron to BS 41:1973 (1981) Specification for cast iron spigot and socket flue or smoke pipes and fittings, or ◆ Mild steel with a min. wall thickness of 3mm, or ◆ Stainless steel with a min. wall thickness of 1mm, to BS 1449: Part 2:1983, or ◆ Vitreous enamelled steel to BS 6999: 1989.	◆ Sheet metal to BS 715: 1989, or ◆ Asbestos cement to BS 567: 1973 (1989) Specification for asbestos cement flue pipes and fittings, light quality, or BS 835:1973, or ◆ Cast iron to BS 41: 1973 (1981) Specification for case iron spigot and socket flue or smoke pipes and fittings, or ◆ Any material described in column "A" of this table.

FLUE PIPES
continued.

Shielding of flue pipes – solid fuel appliances

Where flue pipes are used they should be shielded from adjoining materials as illustrated here.

Flue pipes from solid fuel appliances should be separated from adjoining materials by at least the distances illustrated below.

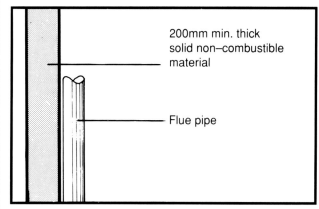

200mm thick solid non–combustible material. There is no distance requirement between the flue pipe and non–combustible material, or,

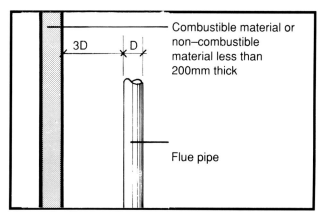

Flue pipe separated from adjoining material by at least 3 times pipe diameter **D**, or,

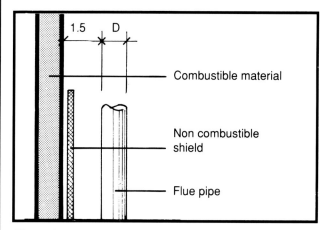

Flue pipe separated from combustible material by a non-combustible shield. There should be a 12.5mm air gap between the shield and combustible material. The width of the shield should be at least three times pipe diameter **D**.

FLUE PIPES
continued.

Shielding of flue pipes – gas burning appliances

Where flue pipes from gas burning appliances pass through a wall, floor or roof, they should be separated from any combustible material by a non–combustible sleeve enclosing an air space of at least 25mm around the flue pipe as illustrated below, or be at least 25mm from any combustible material as illustrated opposite.

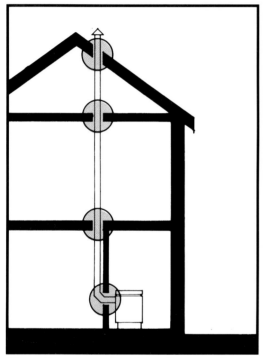

Flue pipe separated from floor, wall and roof by non-combustible sleeve.

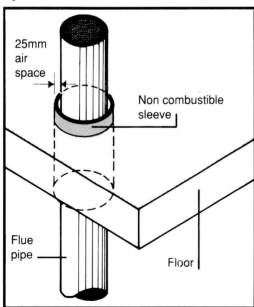

Floor protected by non-combustible shield.

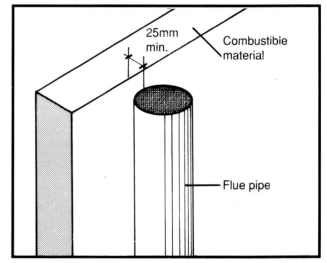

Separation of flue pipe to gas burning appliance from combustible material.

Note: for a double–walled flue pipe, the 25mm distance may be measured from the outside of the inner pipe, as illustrated below.

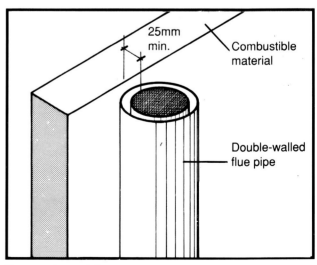

Separation of flue pipe to gas burning appliance from combustible material.

FLUE PIPES
continued.

Sizes of flue pipes

Solid fuel appliance:

Generally for solid fuel appliances a 200mm diameter flue or square section flue of equivalent area (e.g. 175 x 175mm) is required.

Where a flue serves a closed solid fuel appliance and the flue run incorporates an offset, the size of the flue should be increased to 225mm diameter or 200 x 200mm square section flue.

Gas burning appliance:

For gas burning appliances, generally the flue should have a cross–sectional area of at least that of the appliance outlet. In the case of gas fires a round flue of at least 125mm diameter should be used or a rectangular flue of at least 16,500mm^2 cross–section with a minimum dimension of 90mm (e.g. 90 x 185mm approx. or 130 x 130mm approx.).

Oil burning appliances:

For oil burning appliances discharging into a flue pipe the flue size should be at least that of the appliance outlet. Where oil burning appliances connect to chimneys, minimum flue size varies with appliance outlet as outlined in the table below.

Rated output of appliance	Minimum flue size
up to 20kW	100mm dia.
20 – 32kW	125mm dia.
32 – 45kW	150mm dia.

FLUE PIPES
continued.

Flue outlets

In the case of solid fuel appliances, the guidance given on pages 106 to 109 of this manual should be consulted. This guidance also applies to oil burning appliances other than those fitted with balanced flues or pressure jet appliances. Flue outlets to pressure jet appliances should terminate above the roofline.

In the case of balanced flue gas or oil burning appliances the outlet should be:

◆ so located as to allow free air intake and dispersal of flue gases,

◆ located at least 300mm from any wall opening in the case of gas appliances and at least 600mm in the case of oil appliances,

◆ fitted with a guard if the outlet is located where it might come into contact with persons or be subject to damage, and

◆ designed to prevent anything entering the flue which might give rise to restriction or blockage.

Where gas burning appliances are connected to conventional flues, outlets from such flues should be located as follows:

◆ Location should allow free passage of air across outlet.

◆ Outlet to be 600mm minimum from any opening.

◆ Terminal to be fitted if flue is less than 175mm across (unless the flue serves a gas fire).

Direction of flues

Whenever possible flue runs should be vertical. Horizontal runs should be avoided except where appliances have back outlets in which case the horizontal run should not exceed 150mm. Any bend or offset in a flue should be as illustrated below.

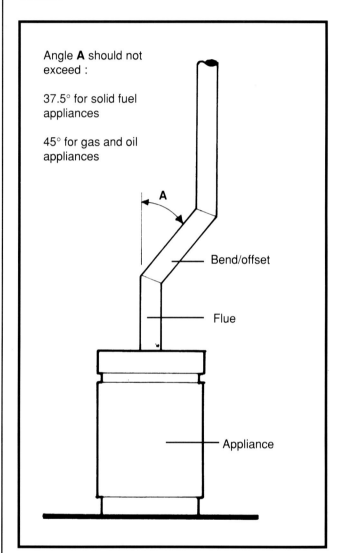

Angle **A** should not exceed :

37.5° for solid fuel appliances

45° for gas and oil appliances

Bend/offset

Flue

Appliance

Direction of flue

HEARTHS

For solid fuel appliances, the hearth should be solid, non combustible and at least 125mm thick (this can include the thickness of a solid non–combustible floor under the hearth).

Combustible material in proximity to a hearth should be as illustrated opposite.

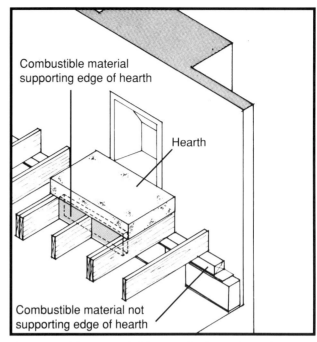

Combustible material supporting edge of hearth

Hearth

Combustible material not supporting edge of hearth

Combustible material in proximity to hearth

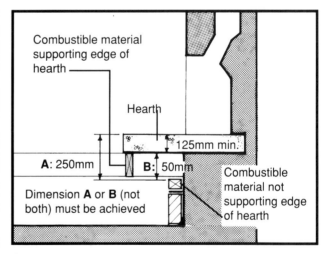

Combustible material supporting edge of hearth

Hearth

125mm min.

A: 250mm   B: 50mm

Dimension **A** or **B** (not both) must be achieved

Combustible material not supporting edge of hearth

Combustible material in proximity to hearth. Combustible material not supporting edge of hearth should be located as to achieve dimension **A** or **B** (not both)

HEARTHS
continued

Solid fuel appliance

Solid fuel appliances placed on hearths should meet the provisions of the illustration below.

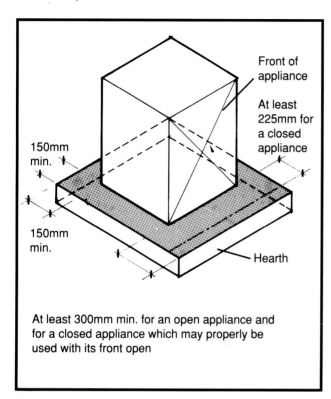

Solid fuel appliance – hearth requirements.

Gas burning appliance

In the case of gas fired appliances (other than solid fuel effect appliances) a hearth as illustrated opposite is required if any flame or incandescent material in the appliance is within 225mm of the floor, or if the appliances do not comply with one of the following standards: Irish Standard 280, 281, 282, 285, 803, or 805, or British Standard 5258, 5386.

In the case of solid fuel effect gas appliances a hearth complying with the requirements for solid fuel appliances should be provided.

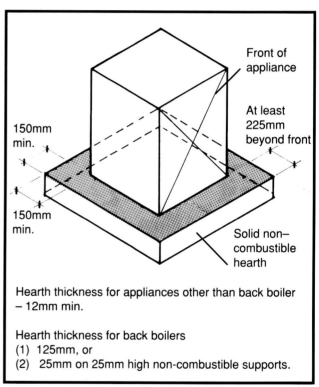

Gas burning appliance – hearth requirements

Oil burning appliances

For oil burning appliances a hearth similar to that for a solid fuel appliance is required where the floor under the appliance is likely to have a surface temperature of 100°C. Otherwise the appliance may stand on a rigid, imperforate sheet of non-combustible material.

SHIELDING OF APPLIANCES

Solid fuel appliances

Solid fuel appliances should be shielded from adjoining combustible material as illustrated below.

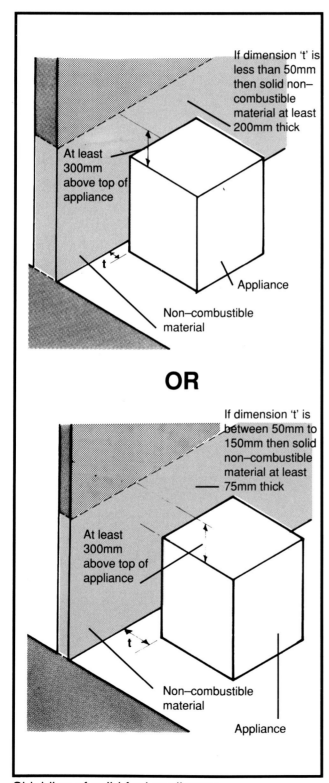

OR

Shielding of solid fuel appliances

Gas burning and oil burning appliances

Illustrated below are the recommendations for the shielding of gas–burning and oil burning appliances.

The recommendations apply to gas burning appliances unless they comply with the relevant provisions of the following standards: Irish Standards 280, 281, 282, 285, 644, 645, 803, 805 and British Standards 5258 and 5386. Similarly, the recommendations apply to oil appliances if the surface temperature of the sides or back of the appliance is likely to exceed 100°C.

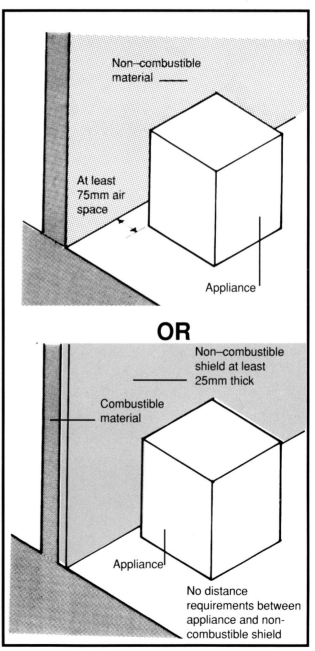

OR

Shielding of gas burning and oil burning appliances

INTRODUCTION

The guidance in this section is reproduced from the second publication by HomeBond about the Building Regulations. It discusses the Regulations with respect to thermal insulation of dwellings, and provides:

♦ Typical construction details for roofs, external walls, windows, and floors;

♦ Discussion of hot water systems, hot water storage vessels, pipes and ducts;

♦ An introduction to thermal insulation concepts;

♦ An outline of the different heat loss calculation methods in the TGD to Part L;

♦ Worked examples of heat loss calculations, to comply with the regulations, for:
- a detached bungalow
- a two-storey semi-detached house
- a two-storey terraced townhouse.

See page 71 for HomeBond insulation requirements where outer leaf of cavity wall is of brickwork or fair faced blockwork construction.

THE BUILDING REGULATIONS, 1991, PART L

Part L of the First Schedule to the Building Regulations, 1991, 'Conservation of fuel and energy', requires (Regulation L1) that "**A building shall be so designed and constructed as to secure, insofar as is reasonably practicable, the conservation of fuel and energy**". Heated areas in all new dwellings, house extensions, and buildings changed to dwelling use, must meet this requirement since 1 June 1992. The requirement may be complied with in different ways. Where work is carried out in accordance with the Technical Guidance Document to Part L, this will, prima facie, indicate compliance with the requirements of the Regulations.

THE TECHNICAL GUIDANCE DOCUMENT TO PART L

The Technical Guidance Document (TGD) expands on the general requirements for energy conservation.

Conservation of fuel and energy in a building requires the provision of energy efficient measures to

♦ limit heat loss and, where appropriate, maximise heat gain through the building fabric

♦ control the output of the space heating and hot water systems

♦ limit heat loss from hot water storage vessels, pipes and ducts.

The Regulations have implications for house construction in each of these areas: building fabric, heating systems, and pipes and storage. The TGD also refers to unheated areas and to extensions:

Unheated ancillary areas such as unheated conservatories, porches, garages and the like do not require specific provisions for the conservation of fuel and energy - but the house to which they are attached must comply with the Regulations.

For extensions not exceeding 6.5m² in floor area, reasonable provision for the conservation of fuel and energy is considered to have been made if the construction is similar to the existing construction.

The TGD to Part L gives three ways to evaluate thermal performance of the building fabric: the Overall, Elemental, and Net Heat Loss Methods.

All three methods are suitable for use with new dwellings. The Elemental Heat Loss Method may also be used for extensions, material alterations and changes of use.

Guidance in the TGD with respect to the use of a particular material, method of construction, standard or specification does not preclude the use of any other suitable material, method of construction, standard or specification.

You should read this appendix with a copy of the TGD beside you for reference.

TYPICAL FLOOR CONSTRUCTIONS

Required insulation thickness for floor varies with:

◆ Conductivity of insulation
◆ Floor area and profile
◆ Number of exposed surfaces

See TGD, Table 9, page 21

If insulation is expanded polystyrene of conductivity **0.037** W/mK, minimum **53** thickness will satisfy the Regulations for most houses.

If insulation is extruded polystyrene of conductivity **0.028** W/mK, minimum **40** thickness will satisfy the Regulations for most houses.

If insulation is positioned above the slab and below a screed:

◆ screed to be at least 65 thick
◆ of adequate strength, e.g., 1:3 mix
◆ take care to avoid cracking: thicken or reinforce screed if necessary

Insulation thicknesses given are those required to satisfy requirements of the Elemental Heat loss Method.
Less insulation may be acceptable if alternative calculation methods are used.
See the advice later in this appendix

Required insulation thickness for suspended floor varies with

◆ Conductivity of insulation
◆ Floor area and profile

See TGD. L, Table 10, page 21

If insulation is expanded polystyrene of conductivity **0.037** W/mK, **53** thickness will satisfy the Regulations.

If insulation is glass fibre quilt of conductivity **0.040** W/mK, **57** thickness will satisfy the Regulations.

Underfloor ventilation:

provide openings at least equal to 1500 mm² clear ope for every 1 m run of wall.

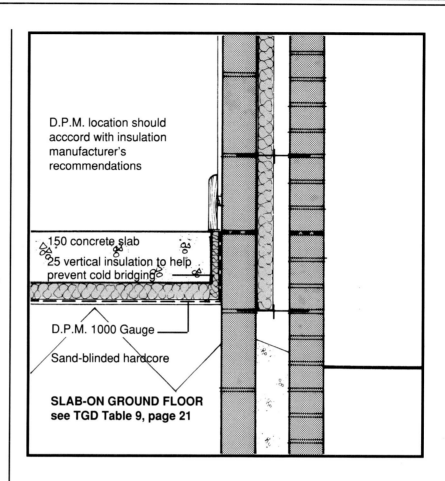

D.P.M. location should acccord with insulation manufacturer's recommendations

150 concrete slab

25 vertical insulation to help prevent cold bridging

D.P.M. 1000 Gauge

Sand-blinded hardcore

SLAB-ON GROUND FLOOR see TGD Table 9, page 21

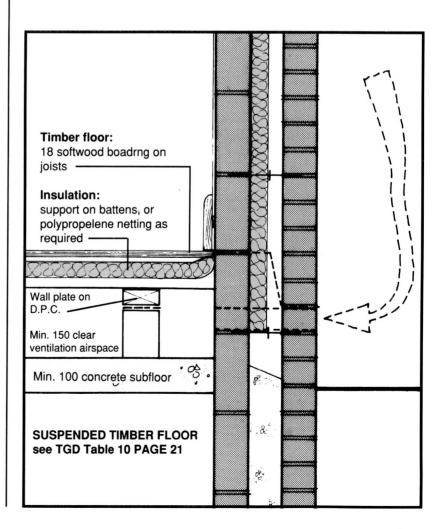

Timber floor:
18 softwood boadrng on joists

Insulation:
support on battens, or polypropelene netting as required

Wall plate on D.P.C.

Min. 150 clear ventilation airspace

Min. 100 concrete subfloor

SUSPENDED TIMBER FLOOR see TGD Table 10 PAGE 21

TYPICAL CAVITY WALL CONSTRUCTION

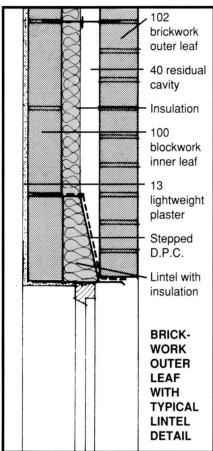

- 102 brickwork outer leaf
- 40 residual cavity
- Insulation
- 100 blockwork inner leaf
- 13 lightweight plaster
- Stepped D.P.C.
- Lintel with insulation

BRICK-WORK OUTER LEAF WITH TYPICAL LINTEL DETAIL

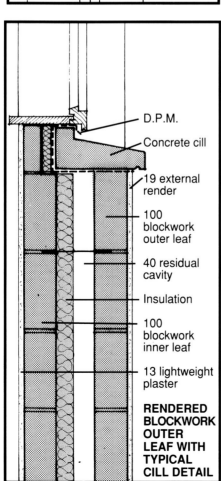

- D.P.M.
- Concrete cill
- 19 external render
- 100 blockwork outer leaf
- 40 residual cavity
- Insulation
- 100 blockwork inner leaf
- 13 lightweight plaster

RENDERED BLOCKWORK OUTER LEAF WITH TYPICAL CILL DETAIL

PLASTERED INTERNALLY

Elemental Heat Loss Method: Walls

0.45 W/m²K is the maximum elemental value for exposed walls.

Brickwork outer leaf with construction shown:
See TGD Table 7, page 18, reference W2

Insulation material	Conductivity	Min. Thickness
Polyurethane board	0.022	35mm
Extruded polystyrene board	0.025	40mm
Glass fibre batt	0.032	50mm
Expanded polystyrene board	0.037	59mm
Glass fibre quilt*	0.040	64mm

At heads, jambs, cill: avoid cold bridging at opes.
U-value to be 1.2 W/m²K or better: 19mm of expanded polystyrene of conductivity 0.037 is required.

Insulation location to be in cavity [as on diagrams] or as internal lining, as manufacturers recommend
* glass fibre quilt fixed to inside face of inner leaf, below plaster.

CAVITY WALL TIE SPACINGS
[from TGD-A, table C3, maximum spacing of cavity ties]

For cavity between 76 and 110 wide,
- horizontal spacing to be 750 maximum
- vertical spacing to be 450 maximum
[or such spacing as will maintain same no. of ties per square metre]
300 maximum vertical spacing for ties at unbonded jambs.
Use vertical twist type ties to I.S. 268:1986, or wire of equal strength.

ELEMENTAL HEAT LOSS METHOD: WALLS

0.45 W/m²K is the maximum elemental value for exposed walls.

Rendered blockwork outer leaf with construction shown:
See TGD Table 7, page 18, reference W1

Insulation material	Conductivity	Min. Thickness
Polyurethane board	0.022	35mm
Extruded polystyrene board	0.025	40mm
Glass fibre batt	0.032	50mm
Expanded polystyrene board	0.037	60mm
Glass fibre quilt*	0.040	65mm

* Glass fibre quilt fixed to inside face of inner leaf, below plaster

TYPICAL CAVITY WALL CONSTRUCTION

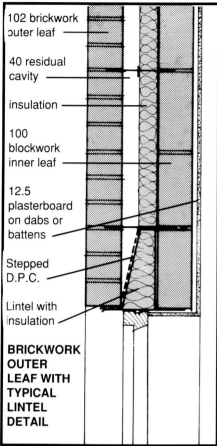

102 brickwork outer leaf

40 residual cavity

insulation

100 blockwork inner leaf

12.5 plasterboard on dabs or battens

Stepped D.P.C.

Lintel with insulation

BRICKWORK OUTER LEAF WITH TYPICAL LINTEL DETAIL

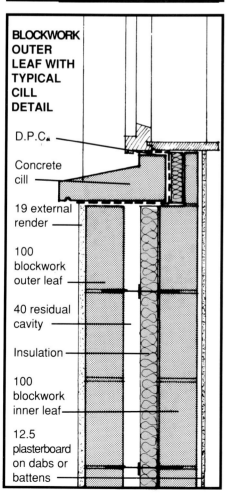

BLOCKWORK OUTER LEAF WITH TYPICAL CILL DETAIL

D.P.C.

Concrete cill

19 external render

100 blockwork outer leaf

40 residual cavity

Insulation

100 blockwork inner leaf

12.5 plasterboard on dabs or battens

DRYLINED INTERNALLY

Elemental Heat Loss Method: Walls

0.45 W/m²K is the maximum elemental value for exposed walls.

Brickwork outer leaf with drylined construction shown:
See TGD Table 7, page 18, reference W4

Insulation material	Conductivity	Min. Thickness
Polyurethane board	0.022	32mm
Extruded polystyrene board	0.025	36mm
Glass fibre batt	0.032	48mm
Expanded polystyrene board	0.037	55mm
Glass fibre quilt*	0.040	64mm

At heads, jambs, cill: avoid cold bridging at opes.
U- value to be 1.2 W/m²K or better: 19mm of expanded polystyrene of conductivity 0.037 is required.

Insulation location to be in cavity [as on diagram] or as internal lining, as manufacturers recommend
*glass fibre quilt fixed to inside face of inner leaf, below plaster.

Cavity wall tie spacings
[from TGD-A, table C3, maximum spacing of cavity ties]

For cavity between 76 and 110 wide,
- horizontal spacing to be 750 maximum
- vertical spacing to be 450 maximum
[or such spacing as will maintain same no. of ties per square metre]
300 maximum vertical spacing for ties at unbonded jambs.
Use vertical twist type ties to I.S.268:1986, or wire of equal strength.

Elemental Heat Loss Method: Walls 0.45 W/m²K is the maximum elemental value for exposed walls.

Rendered blockwork outer leaf with construction shown:
See TGD Table 7, page 18, reference W3

Insulation material	Conductivity	Min. Thickness
Polyurethane board	0.022	33mm
Extruded polystyrene board	0.025	36mm
Glass fibre batt	0.032	48mm
Expanded polystyrene board	0.037	57mm
Glass fibre quilt*	0.040	65mm

TYPICAL CONSTRUCTION AT JAMBS
Cavity and Hollow Block Walls

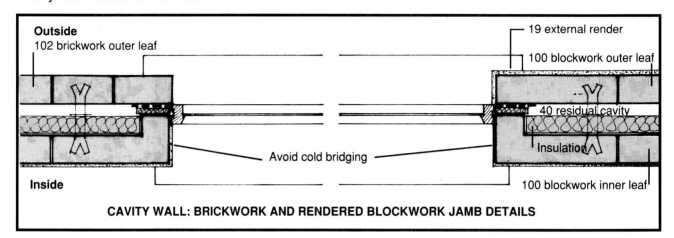

Outside
102 brickwork outer leaf

19 external render

100 blockwork outer leaf

40 residual cavity

Insulation

Avoid cold bridging

Inside

100 blockwork inner leaf

CAVITY WALL: BRICKWORK AND RENDERED BLOCKWORK JAMB DETAILS

All External Opes
Avoid cold bridging.
U-value to be 1.2 W/m²K or better.
See advice page 367 about condensation
and cold bridging.

Material	Conductivity	Min. Thickness
Polyurethane board	0.022	12mm
Extruded polystyrene board	0.025	14mm
Glass fibre batt	0.032	17mm
Expanded polystyrene board	0.037	19mm
Glass fibre quilt	0.040	21mm

Cavity Wall External Opes

Vertical DPC overlaps cavity closer insulation.
Cavity closer insulation overlaps main insulation.

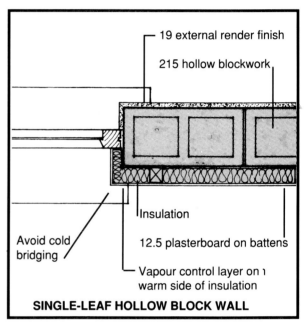

19 external render finish

215 hollow blockwork

Insulation

Avoid cold
bridging

12.5 plasterboard on battens

Vapour control layer on ʋ
warm side of insulation

SINGLE-LEAF HOLLOW BLOCK WALL

If insulation is fixed so as to retain
a cavity between it and the
blockwork, insulation thickness
may be slightly reduced.

If 9.5 thick drylining board is used
in lieu of 12.5, 1mm additional
insulation thickness is required.

ELEMENTAL HEAT LOSS METHOD: WALLS

**0.45 W/m²K is the maximum elemental value for
exposed walls.**

To achieve a U-value of 0.45 W/m²K in the single-leaf hollow
block wall shown:
See TGD Table 7, page 18, reference W6

Insulation material	Conductivity	Min. Thickness
Phenolic foam board	0.018	31mm
Polyurethane board	0.022	38mm
Extruded polystyrene board	0.025	43mm
High density exp. polystyrene	0.032	54mm
Glass fibre batt	0.034	58mm
Expanded polystyrene board	0.037	64mm
Glass fibre quilt	0.040	69mm

TYPICAL ROOF CONSTRUCTION

Provide continuous eaves ventilation in accordance with TGD-F, Ventilation: Ope at least equal to continuous strip 10mm wide

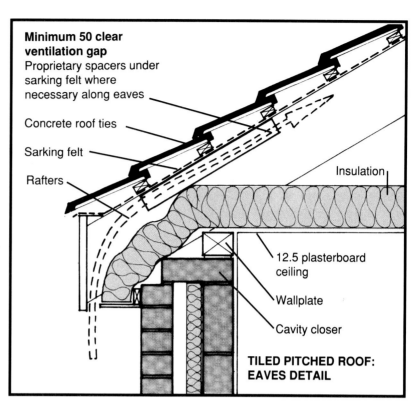

Minimum 50 clear ventilation gap
Proprietary spacers under sarking felt where necessary along eaves

Concrete roof ties

Sarking felt

Rafters

Insulation

12.5 plasterboard ceiling

Wallplate

Cavity closer

TILED PITCHED ROOF: EAVES DETAIL

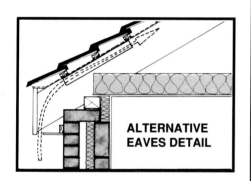

ALTERNATIVE EAVES DETAIL

ELEMENTAL HEAT LOSS METHOD: ROOFS
0.25 W/m²K is the maximum elemental value for exposed roofs.
Tiled pitch roof with construction shown:
See TGD Table 7, page 21, reference R1

Insulation material	Conductivity	Min. Thickness
Polyurethane board	0.022	77mm
Extruded polystyrene board	0.025	87mm
Expanded polystyrene board	0.037	129mm
Glass fibre quilt	0.040	139mm

Where ceiling follows roof slope, batten out rafters so as to achieve enough depth for
♦ Required thickness of insulation
♦ Minimum 50 clear airspace below felt underlay

Note ventilation requirements of TGD-F.

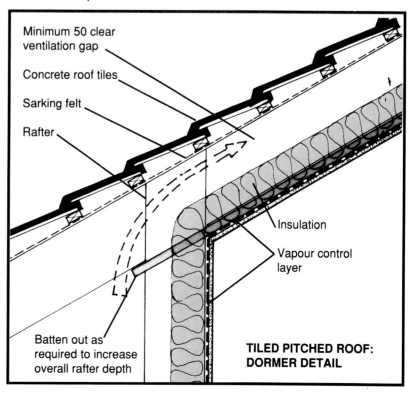

Minimum 50 clear ventilation gap

Concrete roof tiles

Sarking felt

Rafter

Insulation

Vapour control layer

Batten out as required to increase overall rafter depth

TILED PITCHED ROOF: DORMER DETAIL

CONTROL FOR SPACE HEATING AND HOT WATER

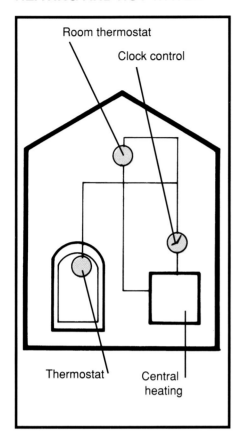

Room thermostat

Clock control

Thermostat

Central heating

INSULATING STORAGE VESSELS, PIPES AND DUCTS

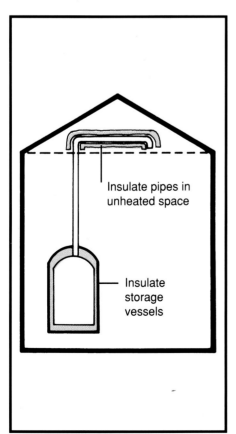

Insulate pipes in unheated space

Insulate storage vessels

The second part of the TGD deals with controls for space heating and for hot water heating. The following are the requirements for dwellings.

CENTRAL HEATING
If there is central heating capable of on-off operation, you should provide a clock control to allow intermittent operation.

and
to control the heating output, you must provide either

◆ A room thermostat, or
◆ Thermostatic radiator valves, or
◆ Any other equivalent from of sensing device.

HOT WATER STORAGE
Unless the hot water is from a solid fuel appliance, all hot water storage vessels must have a thermostat control.

Unless the hot water is from a solid fuel appliance or heated by off-peak electricity, all hot water storage vessels larger than 150 litres must have a time switch to cut off heat when there is no demand.

The third part of the TGD deals with the insulation of hot water storage cylinders, and of pipework and ductwork. The following are the requirements for dwellings.

◆ All hot water storage vessels must be insulated so as to limit their heat loss to 90 W/m² of their surface area.

◆ Factory insulated cylinders may be insulated to BS 699: 1984; BS 1566 Parts 1 and 2: 1984; or, BS 3198;1981 as relevant. Other cylinders can be insulated with a lagging jacket to BS 5615: 1985, with the segments of the jacket taped together to provide an unbroken insulation cover.

◆ Unless a pipe or duct contributes to the useful heat requirement in a room or space, it should be insulated.

- Pipes should be insulated in material with a conductivity value of 0.045 W/mK or better
- Pipes of 40 diameter or less: Insulation thickness to be equal to the pipe outside diameter
- Pipes of over 40 diameter: Insulation thickness: at least 40 mm
- Alternatively, insulate pipes to BS5422:1990
- Insulate ducts to BS 5422: 1990

THERMAL INSULATION CONCEPTS

Conductivity, Resistance, and U-values

Every building material allows to a greater or lesser degree, a flow of heat across it. This measurement is its **thermal conductivity**. Typical conductivities for different materials are given in the TGD, although manufacturers' certified test data should be used in preference. The inverse of conductivity is **thermal resistivity**: resistance to the flow of heat across a material. Some materials, such as metals, or dense concrete, have low thermal resistivity and are poor insulants. Other, such as fiberglass, or polystyrene, have high thermal resistivity and are good insulants.

A table of the thermal conductivity of many materials is given at Table 5, page 13 of the TGD. The following table gives conductivity values provided by the makers of some commonly used insulants. Other insulation products are also available. It is necessary to check the conductivity of each particular insulation material used, because this varies slightly from one manufacturer to another, and the required thickness likewise varies.

Material	Thermal Conductivity, W/mK
Expanded polystyrene board	0.036 to 0.038
Extruded polystyrene board	0.024 to 0.028
Glass fibre quilt	0.034 to 0.040
Glass fibre batt	0.032 to 0.034
Mineral fibre quilt	0.037
Mineral fibre batt	0.032 to 0.033
Phenolic foam board	0.018
Polyurethane board	0.022 to 0.025

Thickness as well as **resistivity** influences overall heat flow through any component. The thicker the material, the less heat flows: 100mm of fiberglass allows less heat through than does 50mm. This combination of thickness and resistivity gives the **overall thermal resistance** of a component: its **R–value**.

In any construction element, such as a roof, an external wall, or a floor, different materials are used in different thicknesses. For example, in an external wall there may be an outer leaf of brickwork; an inner leaf of concrete blockwork, faced with foil–backed plasterboard; a cavity – which contributes to thermal insulation – and a lightweight insulating material in the cavity. The thermal resistance of each component, added together, is the total thermal resistance of the wall. When this resistance figure is inverted, the resultant figure is a measure of how much heat flows from the inside to the outside. This is called the **U–value** of the element.

The TGD sets out certain U–values to be achieved for each exposed or semi-exposed building element in the dwelling: floors; walls; roofs; and external doors and windows.

THERMAL INSULATION CONCEPTS
continued.

◆ **Equivalent thickness of insulation**
When designing a wall or roof, one can consider the thermal resistivity of each element in the construction, work out its thickness, and thereby its resistance; add up the sum, invert the total, and obtain the U–value for the overall construction. This involves a lot of tedious work.
The concept of **equivalent thickness of insulation** allows a shortcut. When designing a roof, for example, start by choosing the type of insulation: say, glass fibre. Read its thermal conductivity from manufacturers' data. Compare this with the required U–values in the TGD, to see how much overall thickness of material, at that conductivity, is needed to comply with the Regulations. All the materials in the roof – the felt, airspace, plasterboard and so on – also help insulate. These other materials, which tend to repeat from building to building in the same order, equal an equivalent thickness of insulation: they equal so much glass fibre in their thermal performance. Table 7, page 18 of the TGD is a list of equivalent thickness of insulation.

◆ **Thin Spacers**
One may ignore the effects of timber joists or framing, battens, thin cavity closures, mortar bedding, damp–proof membranes, metal spacers and other thin separate components, when calculating a U–value.

◆ **Measuring Area and Volume**
The Overall and Net calculation methods involve measuring areas and volume. Rules for measuring are as follows:

To calculate wall, roof and floor areas, and building volumes, measure between internal faces of the appropriate building elements. In the case of roofs, measure in the plane of the insulation.

To calculate window, rooflight and door opening areas, measure between internal faces of appropriate cills, lintels, and reveals. One may count heads, jambs and cills as part of the ope, as preferred.

To calculate volume, count the total volume enclosed by all enclosing elements, including intermediate floors and associated floor spaces.

◆ **Exposed and Semi–Exposed Elements**
The TGD sets out different U–value requirements for exposed, and semi–exposed, elements (floor, wall, roof). Exposed element means an element exposed to the outside air. Semi–exposed element means an element separating any part of a building from an enclosed unheated space which has exposed elements which don't meet the requirements of the Regulations – see diagram 1, page 5 of the TGD.

◆ **Floors**
The amount of insulation needed to meet the required U–value of 0.45W/m^{2}K depends on floor size, shape, the nature of the ground, and whether the floor rests on the ground, or is suspended over it. Floors with longer exposed edges need more insulation than square ones. If the tables in the TGD are used to calculate the amount of insulation needed, houses generally require full underfloor insulation. To calculate the U–value of a floor is complex. The TGD avoids such discussion, and Tables 9 and 10, page 21, show how to meet the requirement.

◆ **Windows and Openings**
Heat losses through windows vary with the number of leaves of glazing, the frame materials, and the construction. In the absence of certified U–values from the window manufacturer, one may use the standard or effective U–values at TGD Table 11, page 22.

Ventilation openings (except windows, rooflights or doors) and meter cupboard recesses may be given the same U–value as the element in which they occur.

One may count lintels, jambs and cills at window, rooflight and door opes either as part of the window, rooflight or door opening area, or as part of the roof, wall or floor in which they occur – as preferred. Cold bridging must be avoided in these locations.

CALCULATION METHOD

The TGD gives three different ways to calculate compliance with Regulation L1 as regards heat losses through the fabric of a house:
(1) The Overall Heat Loss Method; (2) The Elemental Heat Loss Method and (3) the Net Heat Loss Method.

1) THE OVERALL HEAT LOSS METHOD (TGD-L, 1.2)

A **maximum average U-value** of all exposed and semi-exposed elements of the dwelling is allowed. This **maximum average U-value permitted** depends on the ratio between (a) building volume and (b) the total area of exposed and semi-exposed elements. Diagram 1 on page 5 of the TGD illustrates exposed and semi-exposed elements. When calculating the average U-value, notional U-values equal to 0.75 times the actual U-values are used for semi-exposed elements. Calculate building volume, and the total area of the exposed and semi-exposed elements. Depending on the ratio between the two, the maximum average U-value (Um) is set out in Diagram 2 or Table 1, page 7 of the TGD.
The Overall Heat Loss Method also requires maximum average elemental U-value for exposed roofs (0.35W/m²K), exposed ground floors (0.45 W/m²K), and exposed walls (0.55 W/m²K).

◆ To comply with the Regulations by using the Overall Heat Loss Method, the average U-value of the house must equal or better the maximum average U-value permitted by Table 1.
◆ In addition, individual elemental U-values for roofs, walls and ground floors must equal or better those listed.

2) THE ELEMENTAL HEAT LOSS METHOD (TGD-L, 1.3)

This method is suitable for dwellings, material alterations, material changes of use, and extensions.
A maximum elemental U-value in W/m²K is given for each element.

	New buildings and extensions	Material Alterations, Material Changes of Use
Exposed roofs	0.25	0.35
Exposed walls	0.45	0.60
Exposed floors	0.45	0.60
Ground floors	0.45	No requirement
Semi-exposed roofs	0.35	0.60
Semi-exposed walls	0.60	0.60
Semi-exposed floors	0.60	0.60

There are also rules regarding window, rooflight and door openings. Total ope area must be 30% or less of floor area. When the total ope (window, rooflight, external door) area exceeds 20% of floor area, they should all be double glazed. Where ope area is 12% or less of floor area, there is no double glazing requirement.

Window, rooflight and door openings as % of floor area	Minimum proportion of total opes to be double glazed (U-value better than 3.6W/m²K)
20 or over (but not over 30)	100%
Between 12% and 20%	Between 0 and 100%, in proportion
12 or less	No requirement

Solid panels and solid door leaves within an overall window, rooflight or door opening are considered equivalent to double glazing.
◆ To comply with the Regulations by using the Elemental Heat Loss Method, the U-value for each building element must equal or better the elemental U-value in the table.
◆ Glazing must be in accordance with the table.
◆ In new buildings, total area of window, rooflight and door opes must not exceed 30% of floor area.

3) THE NET HEAT LOSS METHOD [TGD-L, 1.4]

This method recognises that, on average, varying amounts of heat are lost through windows facing different ways, as solar gain reduces heat loss. Calculate [a] building volume, [b] total area of exposed and semi-exposed elements, and [c] glazing area and orientation. For semi-exposed elements, use notional U-value equal to 0.75 times the actual U-values. Depending on the ratios between volume, glazing, and exposed element area, a **maximum average effective U-value (U$_{me}$)** is permitted. This changes with window orientation. The figures are at Table 4, page 9 of the TGD.
In addition, maximum average elemental U-values apply. Exposed roofs: (0.35 W/m²K), exposed ground floors (0.45 W/m²K), and exposed walls (0.55 W/m²K).
◆ To comply with the Regulations by the Net Heat Loss Method, the average effective U-value must equal or better the maximum permitted by Table 4, for the particular shape and orientation.
◆ In addition, individual elemental U-values for roofs, walls and ground floors must equal or better those listed.

CALCULATION METHOD
continued

Calculation of U–values
◆ All three methods require calculation of the U–value for each exposed and semi–exposed element: floor, walls, external wall openings, and roof. With standard construction one can refer to tables in the TGD and avoid detailed calculation.
◆ Select the insulation: glass fibre, extruded polystyrene, polyurethane...
◆ Detail the construction: cavity walls, pitched roofs, double glazing,.....
◆ Table 5 on page 13 of the TGD gives the thermal conductivity of common materials, but manufacturers' certified test data should be used in preference. Calculations in this publication use such manufacturers' data.

Slab–on–Ground and Suspended Ground Floors
◆ Work out floor space. Take plan dimensions, decide how many edges are exposed.
◆ If the dwelling is detached: all 4 edges are exposed, and use the actual dimensions.
◆ If it is semi–detached: only 3 edges are exposed. Use floor dimensions (the length of the unexposed side) x (twice the length of the exposed side) – see example 2.
◆ If terraced, only 2 parallel edges are exposed. Use floor dimensions (unexposed side) x (20 metres or over) – see example 3.
◆ Use makers data, or TGD Table 5, page 13 to find out the thermal conductivity of the insulation.
◆ Check either TGD Table 9, page 21 for slab–on–ground, or TGD Table 10, page 21, for suspended floors, for the thickness of insulation needed to meet the required U–value of 0.45 W/m²K.

Exposed Floors, Walls, and Roofs
◆ Refer to maker's data, or TGD Table 5, page 13 for the thermal conductivity of the insulation.
◆ Refer to the basic U–values at TGD Table 7, page 18 for the equivalent insulation thickness of the construction.
◆ Adjust, if filling cavities or drylining voids, by subtracting equivalent thickness at TGD Table 8, page 19.
◆ Refer to TGD Table 6, page 17 for the U–value required and the total thickness needed.

◆ Subtract equivalent thickness from total required thickness to find the thickness of insulation needed.

Semi–exposed Floors, Walls and Roofs
◆ Proceed as for exposed elements. Modify the U–value obtained, if the heat loss calculation methods permits.
◆ This applies to the Overall and Net methods, not to the Elemental method, see example 2.

Windows
◆ See TGD Table 11, page 22 for standard or effective U–value.

General Comments

◆ The simplest calculation method is the Elemental Heat Loss Method. However, if one makes additional calculations, more design flexibility is possible. A compact building has a smaller external surface area and loses less heat than a straggling building of the same volume. The TGD recognises this, so the Overall Heat Loss Method requires less rigorous U–values for compact buildings, because of their more efficient shape. Calculation of surface area and volume permits the use of this method.
◆ Windows can help heat gain by way of solar heating. The TGD recognises this, so that if some heat gain calculations are made by the Net Heat Loss Method, greater design flexibility is allowed.
◆ Calculation of actual U–values for floors is complex and is influenced by soil thermal conductivity and ventilation rates. The TGD does not provide calculation examples, but instead gives thicknesses of insulation to achieve given U–values. An appropriate calculation method may permit more flexibility in complying with the Regulations.

◆ **Required Insulation Thicknesses**

   Tables 6,7 & 8 in the TGD may be used to calculate the required thickness of any insulation material needed to achieve a given U–value. Other calculation methods, including that in paragraph A3 to appendix A of the TGD, may give slightly different thicknesses.
   Calculations in this booklet are based on concrete blocks having a density of 2000 Kg/m³. Other block densities will result in different thermal conductivities.

HEAT LOSS CALCULATIONS

Example 1

The first example is of a simple design.

Ground floor:
Concrete slab-on ground;
55 expanded polystyrene, DPM

External walls:
Two leaves 100 concrete blockwork;
100 cavity,
50 expanded polystyrene-partial fill;
External finish: 19 render;
Internal finish: 13 lightweight plaster

Windows and external doors:
Single glazed, timber frame

Roof:
Pitched tiled; felt; ventilated roofspace;
150 glassfibre quilt at ceiling level
12.5 plasterboard and skim

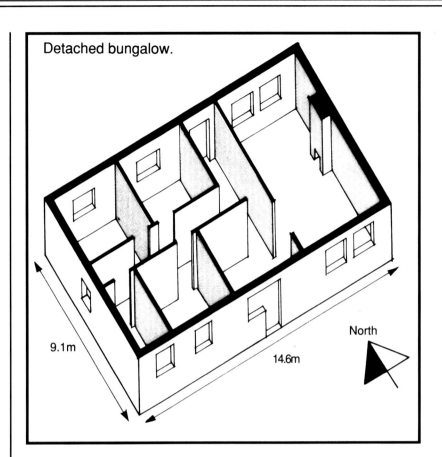

Detached bungalow.

Dimensions

14.6m x 9.1m externally
14.0m x 8.5m internally
2400 floor to ceiling height

Floor area:	**119.00m²**
Volume:	**285.60m³**

External Opes
[Windows, external doors, rooflights]

North:
2 x (1.2x1.0) + (1.2)x(2.1) + 2 x (1.3)x(1.0) = **7.52 m²**
South:
2 x (1.2x1.0) + (0.9)x(2.1) +3 x (1.0)x(1.0) = **7.29 m²**
West:
(0.9x1.0) = **0.90 m²**

Total: **15.71 m²**

Wall Area
(14.0 x 2.4 x 2) + (8.5 x 2.4 x 2) = **108.00 m²**

Example 1
continued

See page 354 for a discussion of how U-values are calculated

ALL METHODS: CALCULATE THE U–VALUES

FLOOR:

Concrete slab-on-ground, 14.0m x 8.5m, exposed all four sides

Insulation: Expanded polystyrene:	Conductivity = 0.037 W/mK: from maker's data
Required U-value	0.45W/m²K
Required insulation thickness for 0.45 W/m²K	53mm, by interpolation from TGD Table 9, page 21
Actual thickness provided	55mm, from above

WALLS:

Insulation: Expanded polystyrene:	Conductivity = 0.037 W/mK: from maker's data
Plastered/rendered cavity blockwork	Equivalent thickness = 21: from TGD Table 7, page 18, W1 No deduction for cavity: this is retained
Insulation thickness	50mm, from the design specification
Total equivalent thickness:	21 + 50 = 71mm
U-value	0.52 W/m²K: by interpolation from TGD Table 6, page 17

WINDOWS:

Single glazed, timber frame:	From the design specification
Standard U-value	5.0 W/m²K: from TGD Table 11, page 22

ROOF:

Insulation type: glassfibre quilt:	Conductivity = 0.040 W/mK from maker's data
Pitched slated; felt; plasterboard	Equivalent thickness = 21: from TGD Table 7, page 18: R1
Insulation thickness	150mm, from the design specification
Total equivalent thickness:	21 + 150 = 171mm
U-value	0.24 W/m²K: from TGD Table 6, page 17

Having calculated the U-values, decide which calculation method to employ. The following two pages show each calculation method applied, in turn, to the design example. It will be seen that a design may be satisfactory using one design method, and unsatisfactory with another.

OVERALL HEAT LOSS METHOD
Example 1
continued.

Calculate area and volume, and average U–value. Floor area, and height to plane of insulation, are as above.

Element- Exposed or semi–exposed	Area m²	U–value W/m²K	Area x U–value W/K
Floor: (14.0 x 8.5)	119.00	0.45	53.55
Wall: (walls minus opes)	92.29	0.52	47.99
Opes: (single glazed, wood frame)	15.71	5.00	78.55
Roof:	119.00	0.24	28.56
	346.00		208.65

Total Area (A$_t$)	346.00m²
Total AU	208.65W/K

The actual average U–value of the dwelling.
$$U_m = \frac{Total\ AU}{A_t} = \frac{208.65}{346.00} = 0.60\ W/m^2K$$

From Diagram 1, TGD page 7, calculate the maximum allowable average U–value, U_m, having regard to the area and volume of the building.

Area of exposed and semi–exposed elements
A_t = 346.00 m²;
Building Volume [V] = 285.60 m³;

$$\frac{A_t}{V} = \frac{346.00}{285.60} = 1.21$$

Maximum allowable U_m [from TGD Table 1 p.7]
= 0.60 W/m²K

On this calculation basis the proposed specifications are acceptable.

Check the maximum average elemental U–values have also been met.

Element	Max. Aver. Elem. U Value	Calculated U–Value
Ground floor	0.45 W/m²K	0.45–satisfactory
Walls	0.55 W/m²K	0.52–satisfactory
Roof	0.35 W/m²K	0.24–satisfactory

If the design is checked by the Overall Heat Loss Method, it is satisfactory.

ELEMENTAL HEAT LOSS METHOD

Having worked out the actual U-value of each element, the Elemental Heat Loss Method involves
1) checking actual U-value against those required and
2) an examination of the area of window, rooflight and door openings, to see (a) what proportion, if any, must be double glazed; and (b) that the total area does not exceed 30% of the floor area.

(1) Checking of the actual U-value against those required

Element	Max. Aver. Elem. U-value	Calculated U-value
Ground floor	0.45 W/m²K	0.45- satisfactory
Exposed walls	0.45 W/m²K	0.52- not satisfactory
Exposed roof	0.25 W/m²K	0.24- satisfactory

To lower the U-value for exposed walls to 0.45 W/m²K, increase the quantity of insulation. To find out by how much, study TGD Table 6, page 17. Conductivity of expanded polystyrene = 0.037 W/mK as before.
Required insulation thickness: 81mm (by interpolation, Table 6, page 17)
Wall construction equivalent thickness: 21 as before, no deduction for cavity: this will be retained
Total insulation: 81-21 = 60mm
Use 60 insulation in the cavities in lieu of the initially proposed 50mm.

(2) Area of openings in the external envelope: windows, rooflights, external doors

Total openings in roof and external walls = 15.71 m², from above; Floor area = 119.00 m², also from above; Total proportion of openings area to floor area: 13.2%

◆ On this basis, 15% of the openings must be double glazed. At present all opes are single glazed.
Double glaze 15% of the openings: 2.3 m² approximately.
Solid panels and solid door leaves within an overall window, rooflight or door opening are considered equivalent to double glazing.
◆ The 13.2% proportion of ope area to floor area satisfies the requirement that this be ≤ 30%.

If the design is checked by the Elemental Heat Loss Method, two changes must be introduced:
[1] thicker wall insulation
[2] some double glazing, or the equivalent area in solid external doors

NET HEAT LOSS METHOD
Example 1
continued

As with the Overall Heat Loss Method, first calculate building area and volume, then the average U–value. Floor area and height to plane of insulation are as above. Glazing orientation is important: the effective U–value varies with orientation: TGD table 11, page 22.
U–value for the windows in the current example are as follows (from TGD, Table 11, page 22):

Glazing type	Orientation	Effective U–value
Single glazing, timber frame:	North:	$U = 4.2$ W/m²K
	South:	$U = 2.4$ W/m²K
	East or west:	$U = 3.6$ W/m²K

Element- Exposed or semi–exposed	Area m²	U–value W/m²K	Area x U–value W/K
Ground floor:	119.00	0.45	53.55
Wall:	92.29	0.52	47.99
(wall minus opes)			
Opes: (single glazed,wood frame)			
-North (entrance)	7.52	4.20	31.58
-West	0.90	3.60	3.24
-South (rear)	7.29	2.40	17.50
Roof:	119.00	0.24	28.56
	346.00		182.42

Total Area (A$_t$)	**346.00m²**
Total AU	**182.42 W/K**

The actual average effective U–value of the dwelling,

$$U_{me} = \frac{\text{Total Au}}{A_t} = \frac{182.42}{346.00} = 0.53 \text{ W/m}^2\text{K}$$

By reference to Table 4, TGD page 9, calculate the maximum allowable average effective U–value, U_{me}, having regard to the fabric heat loss area (A$_t$), building volume (V), and glazing area, (A$_g$).

Area of exposed and semi–exposed elements
A$_t$ = 346.00 m² from above.
Building Volume V = 285.60 m³, from above
Glazing area A$_g$ = 15.71 m², from above.

$$\frac{A_g}{V} = \frac{15.71}{285.60} = 0.055$$

$$\frac{A_t}{V} = \frac{346.00}{285.60} = 1.21$$

Maximum allowable average effective U–value, U_{me} [by interpolation from Table 4, TGD, p. 9] = 0.53 W/m²K, equalling the actual U_{me} of 0.53.

On this calculation basis the proposed specifications are acceptable.

Check the maximum average elemental U–value have also been met.

Element	Max. Aver. Elem. U–value	Calculated U–value
Ground floor	0.45 W/m²K	0.45 – satisfactory
Walls	0.55 W/m²K	0.52 – satisfactory
Roof	0.35 W/m²K	0.24 – satisfactory

If the design is checked by the Net Heat Loss Method, it is satisfactory.

COMMENT

- Common to all three calculation methods is the requirement to calculate the U-values of the elements: walls, roofs, floor. These U-values do not vary if the construction and materials remain the same.
- This explains the advice in general HomeBond leaflet on the Regulations. If one uses:
- standard pitched roof construction with 150 glassfibre of conductivity = 0.040 W/mK
- standard two-leaf plastered/rendered insulated cavity wall construction with 60 expanded polystyrene of conductivity = 0.037 W/mK
- standard ground floor concrete slab on ground with 55 expanded polystyrene of conductivity = 0.037 W/mK
- the construction meets the requirements of the Elemental Heat Loss Method and the area of external opes is the only concern. If these are solid, or double glazed throughout, the only calculation is to check that the area of external opes is less than 30% of floor area.

- The following table shows the three different calculation methods applied to the bungalow:

Method	Regulations require	Actually provided
Overall Heat Loss Method ◆ Complies	U_m of 0.60	U_m of 0.60
Elemental Heat Loss method	U of 0.45 in floor U of 0.45 in walls U of 0.25 in roof 15% double glazing	U of 0.45 U of 0.52 U of 0.24 No double glazing
◆ Had to change wall insulation and install double glazing		
Net Heat Loss Method ◆ Complies	U_{me} of 0.53	U_{me} of 0.53

HEAT LOSS CALCULATIONS
Example 2

This two storey semi-detached house has an integral, unheated, garage. In accordance with diagram 1, page 5 of the TGD, the coloured walls and floor around and over this garage are treated as semi-exposed elements.

Insulate roof at ceiling level.

CONSTRUCTION

Ground floor:
Concrete slab-on-ground
50 extruded polystyrene, DPM

Suspended floor over garage:
22 softwood boarding on 175x44 joists
100 glassfibre matt between joists
15 fibre-rated gypsum board to soffit

External walls:
Single leaf 215 hollow concrete blockwork
60 expanded polystyrene to inner face
External finish: 19 render
Internal finish: 12.5 foil-back plasterboard

Windows and external doors:
Double glazed, metal frame

Roof:
Pitched tiled; felt; ventilated roofspace
150 glassfibre quilt at ceiling level
12.5 plasterboard and skim

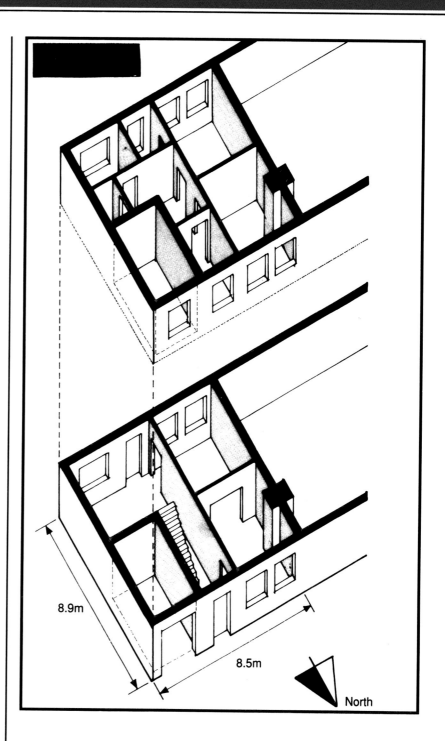

8.9m

8.5m

North

Dimensions

8.5m x 8.9m externally
7.9m x 8.3m internally
2400 floor to ceiling height
200 deep first floor zone
5000 floor to top floor ceiling height

Floor areas: **119 m²** after deduction for unheated garage
Volume: **297 m³** after deduction for unheated garage

External wall opes
North: **10.35 m²**
South: **10.59 m²**

ALL METHODS: CALCULATE THE U–VALUE

◆ As example one.
◆ Work out the floor shape. Here, 3 edges are exposed. Treat the floor as having dimensions (the length of the unexposed side) x (twice the length of the exposed side)
◆ Exclude unheated area and volume from the calculations.

Ground floor:	Insulation: Extruded polystyrene:	Conductivity = 0.028 W/mK: from maker's data
	Concrete slab–on–ground, 5.2x8.3, with one 8.3 edge unexposed	Treat as 8.3x10.4, with 4 exposed edges
	Required U–value	0.45 W/m²K
	Required insulation thickness	40mm, by interpolation from TGD Table 9, page 21
	Actual thickness provided	50mm, from above: requirement is satisfied.

Floor over garage The specified construction is not detailed in TGD Table 7, page 18. Accordingly the U–value must be calculated, in similar manner to TGD example 1 page 14.

Component	Thermal Conductivity of material: (W/mK)	Material thickness: (m)	Thermal resistance (m²K/w)	Reference
Inside surface			0.14	TGD, para. A2, p. 13
Timber boarding	0.14	0.022	0.15	
Glass fibre quilt	0.04	0.100	2.50	
Gypsum board	0.16	0.015	0.09	
Outside surface			0.04	TGD, para. A2, p. 13
Total resistance			**2.92**	

U–value = 1/2.92 = 0.34 W/m²K
In accordance with TGD para. 1. 2, page 6 for overall and net calculations, a notional U–value of 0.75 times the actual U–value should be used for semi–exposed elements.
Accordingly:
Notional U–value = (0.75) x (0.34) = 0.25 W/m²K

Walls:

Insulation: Expanded polystyrene:	Conductivity = 0.037 W/mK, from manufacturer's data
Plastered/rendered hollow blockwork	Equivalent thickness = 21, from TGD Table 7, p. 18, W6
Dry lining cavity is being filled	Deduct 4mm for cavity to be insulated
Equivalent thickness of wall	21 – 4 = 17mm, see TGD table 8, page 19
Insulation thickness	60mm, from the design specification
Total equivalent thickness:	17 + 60 = 77mm
U–value	0.48 W/m²K, from TGD Table 6, page 17, by interpolation

Walls to Garage

Use notional U–value of 0.75 x (Actual U–value)– TGD, para. 1.2, page 6. 0.75 x 0.48 = 0.36
Notional U–value   0.36 W/m²K

Windows:

Double glazed, metal frame: From the design specification
Standard U–value   3.3 W/m²K, from TGD Table 11, page 22

Roof:

Insulation type: glassfibre quilt:	Conductivity = 0.040 W/mK, from manufacturer's data
Pitched slated; felt; plasterboard	Equivalent thickness = 21, from TGD Table 7,page 18: R1
Insulation thickness	150mm, from the design specification
Total equivalent thickness:	21 + 150 = 171mm
U–value	0.24 W/m²K, from TGD Table 6, page 17

OVERALL HEAT LOSS METHOD
Example 2
continued

Element	Area	U–value	Area x U–value
Exposed or semi–exposed	m²	W/m²K	W/K
Ground floor, excl. garage	53.70	0.45	24.17
Semi–exposed floor over garage	11.88	0.25	2.97
Exposed wall	82.52	0.48	39.61

(Total external wall to heated volume, less external opes and less semi–exposed wall to garage)

Semi–exposed wall to garage	17.04	0.36	6.13
Windows:	20.94	3.30	69.10
double glazed, metal frame, doors			
Roof	65.57	0.24	15.74
	251.65		157.72

Total Area (A$_t$)	251.65 m²
Total AU	157.72 W/K

Actual average U–value of dwelling,
$$U_m = \frac{Total\ AU}{A_t} = \frac{157.72}{251.65} = 0.63\ W/m^2K$$

From Diagram 1, TGD page 7, calculate the maximum allowable average U–value, U$_m$, having regard to the area and volume of the building. Area of exposed and semi–exposed elements:
A$_t$ = 251.65 m²;
Building Volume [V] = 297.00m³ from above
$$\frac{A_t}{V} = \frac{251.65}{297.00} = 0.85$$

Maximum allowable Um [from TGD Table 1, page 7] = 0.67 W/m²K
On this calculation basis the proposed specifications are acceptable.

Now check the maximum average elemental U–values have also been met.

Element	Max. Aver. Elem. U–value	Calculated U–value
Ground floor	0.45 W/m²K	0.45 – satisfactory
Walls	0.55 W/m²K	0.48 – satisfactory
Roof	0.35 W/m²K	0.24 – satisfactory

The design as checked by the Overall Heat Loss Method is satisfactory.

ELEMENTAL HEAT LOSS METHOD

(1) Check the actual U–values against those required.

Element	Max. Aver. Elem. U–value	Calculated U–value
Ground floor	0.45 W/m²K	0.45 – satisfactory
Semi–exposed floor	0.60 W/m²K	0.34 – satisfactory
Exposed walls	0.45 W/m²K	0.48 – not satisfactory
Semi–exposed walls	0.60 W/m²K	0.48 – satisfactory
Exposed roof	0.25 W/m²K	0.24 – satisfactory

To bring the U–value for exposed walls down to 0.45 W/m²K, increase the insulation: TGD Table 6, page 17.

Conductivity for expanded polystyrene = 0.037 W/mK, as before.

Required thickness of insulation is 82mm, by interpolation on the table.

Wall construction equivalent thickness = 17mm as before, deduct for cavity because this will be filled.

Total required insulation thickness = 82 – 17 or 65mm in lieu of the initially proposed 60mm.

(2) Examination of the amount of glazing

Total area of openings in external wall = 20.94m², from above;

Total Floor area = 119.00 m², also from above;

Total proportion of openings area to floor area: 18%.

On this basis, 75% of the openings must be double glazed. This is provided.

The 18% proportion of glazed area to floor area satisfies the requirement that this be ≤30%.

NET HEAT LOSS METHOD
Example 2
continued.

Element-	Area	U-value	Area x U-value
Exposed or semi-exposed	m²	W/m²K	W/K
Ground floor excl. garage as before	53.70	0.45	24.17
First floor semi exposed over garage as before	11.88	0.25	2.97
Exposed wall as before	82.52	0.48	39.61
Semi exposed wall to garage as before	17.04	0.36	6.13
Opes: (double glazed, metal frame)			
North elevation	10.35	2.50	25.88
South elevation	10.59	0.90	9.53
Roof:	65.57	0.24	15.74
	251.65		124.03

Total Area (A_t)   251.65m²
Total AU   124.30 W/K

The actual U-value of the dwelling,

$$U_m = \frac{\text{Total AU}}{A_t} = \frac{124.30}{251.65} = 0.49 \text{ W/m}^2\text{K}$$

Area of exposed and semi-exposed elements
$A_t = 251.65 \text{ m}^2$;
Building Volume [V] = 297.00m³ from above.
Glazing area A_g = 20.94m², from above.

$$\frac{A_g}{V} = \frac{20.94}{297.00} = 0.070$$

$$\frac{A_t}{V} = \frac{251.65}{297.00} = 0.8473$$

Maximum allowable average effective U–value, U_{me} [table 4, TGD p. 9] = 0.55 W/m²K.
On this calculation basis the specification are acceptable.

Check the maximum average elemental U–values have also been met.

Element	Max. Aver. Elem. U–Value	Calculated U–Value
Ground floor	0.45 W/m²K	0.45–satisfactory
Walls	0.55 W/m²K	0.48–satisfactory
Roof	0.35 W/m²K	0.24–satisfactory

If the design is checked by the the Net Heat Loss Method, it is satisfactory.

Comment

◆ Note as with example one, that if the Elemental Heat Loss method is used the initial specifications do not comply, whereas if either the Overall or Net heat Loss Methods are used, the initial specifications do comply.

◆ The required U–values in the walls around the garage are lower than those in the general external walls it would be possible to reduce the insulation thickness in those walls and yet comply with the regulations.

HEAT LOSS CALCULATIONS
Example 3

Example three features a two story mid–terrace house with south facing patio doors.
When examined under both the Overall and the Elemental Heat Loss Methods this house as designed, would not comply with the Regulations.

When examined under the Net Heat Loss Method, it does comply.

CONSTRUCTION

Ground floor:

Suspended timber:
18 softwood boarding
50 glassfibre quilt between joists

External walls:
102 brick outer leaf
110 cavity
60 glass fibre batt–partial fill
100 concrete block inner leaf
13 gypsum plaster

Windows and external doors:
Single glazed, timber frame

Roof:
Pitched tiled; felt;
ventilated roofspace
150 mineral fibre mat, ceiling level
12.5 plasterboard

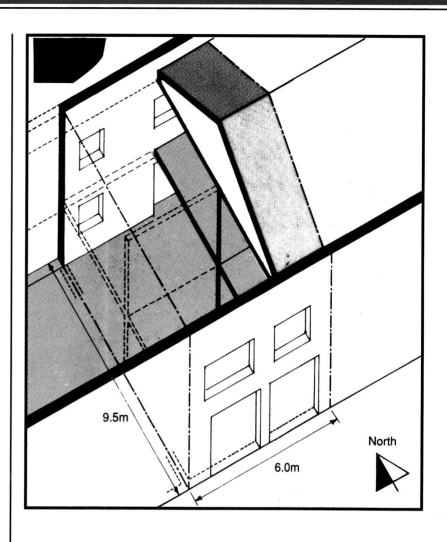

9.5m

North

6.0m

DIMENSIONS
9.0m x 5.8m internally
2600 floor to ceiling height
200 deep first floor zone
5400 high insulated space

Total floor area	**104.40m²**
Volume	**281.88m³**

External wall area
5.4 x 5.8 x 2 =   62.64m²

External opes
Ground floor:
South: (2.0 x 2.1) + (2.0 x 2.1)   =   8.40
North: (0.9 x 2.1) + (1.0 x 1.0)   =   2.89

First floor:
South: (1.5 x 1.2) + (2.0 x 1.2)   =   4.20
North: (1.0 x 1.0) + (1.0 x 1.0)   =   2.00

Total south	**=**	**12.60m²**
Total north	**=**	**4.89m²**
Total	**=**	**17.49m²**

Example 3
continued.

ALL METHODS: CALCULATE THE U–VALUES

GROUND FLOOR:

Suspended timber, size 9.0m x 5.8m, 2 unexposed edges.
Treat as 20+m x 9.0m, 4 exposed edges.

Insulation: glassfibre mat:	Conductivity = 0.040 W/mK: from manufacturer's data
Required U value	0.45 W/m²K
Required insulation thickness	40mm, from TGD Table 10, p.21
Actual thickness provided	50mm, from the design specification

WALLS:

Insulation: glass fibre batt:	Conductivity=0.032 W/mK, from manufacturer's data
Plastered blocks, brick facings externally	Equivalent thickness = 20, from TGD Table 7, p.18, W2.
Partial cavity fill insulation	60mm, from the design specification
Total equivalent thickness:	20 + 60 = 80mm
U-value	0.40 W/m²K, from TGD Table 6, p. 17 by interpolation

WINDOW:

Single glazed, timber frame:	From the design specification
Standard U value	5.0W/m²K, from TGD Table 11, p.22

ROOF:

Insulation type: mineral fibre quilt:	Conductivity = 0.037 W/mK, from manufacturer's data
Pitched slated: felt; plasterboard	Equivalent thickness = 19, from TGD Table 7, p. 18: R1
Insulation thickness	150, from the design specification
Total equivalent thickness:	19 + 150 = 169mm
U-value	0.23 W/m²K, from TGD Table 6, p. 17 by interpolation

OVERALL HEAT LOSS METHOD
Example 3
continued

Element-	Area	U-value	Area x U-value
Exposed or semi-exposed	m²	W/m²K	W/K
Ground floor: (9.0 x 5.8)	52.20	0.45	23.49
Exposed wall: (walls minus opes)	45.15	0.40	18.06
Opes: (single glazed, wood frame)	17.49	5.00	87.45
Roof;	52.20 167.04	0.23	12.01 141.01

Total Area (A_t)	167.04m²
Total AU	141.01 W/K

The actual average U-value of the dwelling,

$$U_m = \frac{\text{Total AU}}{A_t} = \frac{141.01}{167.04} = 0.84 \text{ W/m}^2\text{K}$$

From Diagram 1, TGD page 7, calculate the maximum allowable average U-value, U_m, having regard to the area and volume of the building. Area of exposed and semi–exposed elements A_t= 167.04 m²; Building Volume [V] = 281.88 m³;

$$\frac{A_t}{V} = \frac{167.04}{281.88} = 0.59$$

Maximum allowable U_m [from TGD Table 1, p. 7] = 0.79 W/m²K

The proposed specification is not acceptable as the actual U_m is 0.84 W/m²K and the maximum allowable, 0.79 W/m²K. Examining the (Area x U-value) figures the main problem is with the windows. Over half the heat loss (87.45 out of 141.01) is through this one element. Examine the external opes to find a number of ways of solving the problem. You already know [1] the maximum allowable U-value (U_m), 0.79, and [2] the area of the exposed and semi-exposed elements of the house (A_t) 167.04. Derived from the formula above, this gives you a total permissible AU as follows:

$$\text{Maximum permissible } U_m = \frac{AU}{167.04} = 0.79;$$

Permissible AU = 0.79 x 167.04;
Permissible AU = 131.96.

If you leave the ground floor (23.49), exposed wall (18.06), and roof (12.01) unchanged, then the maximum AU allowable for the external opes is (131.96)–(53.56) = 78.40 W/K.

One could either:

[1} Reduce the area of external opes: Reduce to 15.0m², and recalculate.
Wall: 47.64 x 0.40 = 19.06
Opes: 15.00 x 5.00 = 75.00
AU = 23.49 + 19.06 + 75.00 + 12.01
AU = 129.56 – acceptable.

[2] Double glaze all the opes:
Permissible external ope AU = 78.40;
U = 2.90 for double glazing;
External ope area unchanged of 17.49 m² gives an actual AU of 2.90 x 17.49 or 50.72 – acceptable.

[3] Double glaze some of the opes by calculating to reach a total AU of 78.40 or less for external opes.

Now check that the maximum average elemental U-values have been met.

Element	Max. Aver. Elem. U Value	Calculated U-Value
Ground floor	0.45 W/m²K	0.45 – satisfactory
Walls	0.55 W/m²K	0.40 – satisfactory
Roof	0.35 W/m²K	0.23 – satisfactory

The design needs revising to comply with the Regulations by calculation using the Overall Heat Loss Method. This may be most easily carried out by redesigning either the size or the construction of the external opes.

Elemental heat loss method

(1) Check the actual U values against these required.

Element	Max. Aver. Elem. U Value	Calculated U-Value
Ground floor	0.45 W/m²K	0.45 – satisfactory
Walls	0.45 W/m²K	0.40 – satisfactory
Roof	0.25 W/m²K	0.23 – satisfactory

(2) Examination of the amount of glazing

Total area of openings in external wall = 17.49 m², from above;
Total floor area = 104.40 m², also from above;
Total proportion of openings area to floor area: 16.75%
According to TGD table 3, p.8, at least 60% of the windows must be double glazed so as to meet the requirements by way of the elemental Heat Loss Method calculation.

NET HEAT LOSS METHOD
Example 3
continued.

Element-	Area	U-value	Area x U-value
Exposed or semi-exposed	m²	W/m²K	W/K
Ground floor: (9.0 x 5.8)	52.20	0.45	23.49
Exposed wall: (walls minus opes)	45.15	0.40	18.06
Opes: (single glazed, wood frame)			
-South elevation	12.60	2.40	30.24
-North elevation	4.89	4.20	20.54
Roof;	52.20	0.23	12.01
	167.04		104.34

Total Area (A$_t$)		**167.04m²**
Total AU		**104.34 W/K**

The actual average U-value of the dwelling,

$$U_m = \frac{\text{Total AU}}{A_t} = \frac{104.34}{167.04} = 0.62 \text{ W/m}^2\text{K}$$

Area of exposed and semi–exposed elements
A_t = 167.04 m²
Building Volume V = 281.88 m³, from above.
Glazing area A_g = 17.49 m², from above.

$$\frac{A_g}{V} = \frac{17.49}{281.88} = 0.062$$

$$\frac{A_t}{V} = \frac{167.04}{281.88} = 0.59$$

Maximum allowable average effective U-value, U_{me} [by extrapolation from Table 4, TGD p. 9] = 0.64 W/m²K.

On this calculation basis the specifications are acceptable without change.

Now check that the maximum average elemental U-values have also been met.

Element	Max. Aver. Elem. U Value	Calculated U-Value
Ground floor	0.45 W/m²K	0.45 – satisfactory
Walls	0.55 W/m²K	0.40 – satisfactory
Roof	0.35 W/m²K	0.23 – satisfactory

If the design is checked by the Net Heat Loss Method, it is satisfactory without change.

Discussion

◆ This house is slightly 'climate conscious' as the bulk of the glazed area faces south. As a result, the calculations under both Overall and Elemental Methods, which do not take account of solar gain, are unsatisfactory. However, if the house is reviewed under the Net Heat Loss Method, the design is satisfactory with the particular south orientation of the bulk of the glazing.

◆ Changing the house orientation will make the house unsatisfactory by the Net Heat Loss Method.

GENERAL COMMENTS

◆ Condensation and cold bridging

The Department of the Environment advise that in general, heat flow perpendicular to the inner surface of the building element is assumed for the U–values specified in the TGD to Part L, and that in particular, the value of 0.9 W/m²K given in Clause 0.13 in relation to lintels, jambs, and cills at the door and window openings relates to heat flow perpendicular to the inner surface of the element in which these openings occur. It should be noted that, depending on their detailed design, there may be significant additional heat flow paths through the return sides of such openings. Particular care should be taken to avoid harmful effects of cold bridging at these points. The Department advise that guidance on appropriate practice can be found in the Building Research Establishment document "Thermal Insulation: Avoiding Risk", BR143, 1989, which is referred to in the TGD.

The BRE document advises that a U-value not exceeding 1.2 W/m²K be achieved around external wall opes, as does BS 5250: 1989: 'Code of practice for the control of condensation in buildings'.

◆ The Regulations Generally

Under Part D of the First Schedule to the Building Regulations, all work must be carried out with proper materials and in a workmanlike manner. 'Proper materials' are defined as those fit for their intended purpose and conditions of use. This includes materials bearing a CE mark in accordance with the Construction Products directive, and those complying with an appropriate Irish standard or Irish Agrément Board Certificate. Many insulation manufacturers have Irish Agrément Board Certificates for their materials. A material should be used only in accordance with the advice on an IAB Certificate (or similar) and in accordance with manufacturers' recommendations, In particular, where composite boards are used, special attention should be paid to mechanical fixing and to placement of cavity barriers.

Measures taken to comply with Part L of the Regulations must not conflict with other requirements, such as pertain to Fire Safety or Ventilation. For example, combustible materials used as cavity linings must accord with the requirements for cavity barriers and surface spread of flame ratings necessitated by Part B. The Irish Agrément Board Certificate for a material list conditions for its proper use in such situations.

◆ Relevant Publications

An extensive list of relevant publications is given on pages 28 and 29 of the TGD, and careful study of Technical Document to Part L is indispensable.
The following may be of particular interest:
Insulation of external walls in housing: An Foras Forbartha, 1987, and
Thermal insulation: Avoiding Risk: Digest 143, Building Research Establishment, 1989.